CW00348141

ISBN 978-0-260-53861-1
PIBN 10524656

" Price — $1.12

[J]ames H. Farnsworth's
Book

Bot. Oct. 4th 1824 of Goodwin

[illegible]

[illegible]

[illegible]

[illegible]

PHILOSOPHICAL INSTRUCTOR:

OR,

WEBSTER'S ELEMENTS

OF

NATURAL PHILOSOPHY,

SUBDIVIDED INTO

PRINCIPLES AND ILLUSTRATIONS.

INTENDED FOR

ACADEMIES, MEDICAL SCHOOLS, AND THE POPULAR CLASS-ROOM.

BY AMOS EATON,

Professor of Natural Philosophy and Chemistry in the Vermont Academy of Medicine, Lecturer in the Troy Lyceum, &c. &c.

ALBANY:

PRINTED BY WEBSTERS AND SKINNERS,

At their Bookstore, in the White House, corner of State and Pearl-streets.

1824.

EDITOR'S PREFACE.

WEBSTER'S Elements of Natural Philosophy was written in the true character of a well proportioned abridgment. He seems to have written his little treatise from a recollection of principles, previously digested and made completely his own.

It should be the object of a concise elementary system of natural philosophy to present a clear view of all the important phenomena appertaining to this department of human knowledge, with such explanations and general solutions as the present state of the science will afford. But those easy applications, which cannot escape any ordinary mind, ought not to encumber such a work, but should be left for teachers and students to supply. Such is Webster's Elements.

Professor Patterson's American edition of this work, published in 1808, is out of print. Had his edition still been in the booksellers' shops, I should not have undertaken this; though I can but hope, that both teacher and pupil will derive considerable advantage from my subdivisions into Principles, Illustrations, and Remarks.

My alterations, though numerous, appeared to me essential, according to the plan of this edition. Besides the alterations, I have added every important improvement and discovery in the science, which comports with the object of the work.

The article on Geometry I prefixed. As nothing is included in this article, but what is essential for studying this treatise, it necessarily consists of unconnected materials, sparingly dealt out to the student who has wholly omitted the study of mathematics.

Some improvements having been made in meteorology since Webster wrote his article on that subject, I have written it wholly over. The short articles on the rainbow, halo, parhelion, galvanism, ignis fatuus, aurora borealis, tides, the application of geometrical trigonometry to astronomical calculations, and remarks on the four new planets and aerolites, I added wholly.

This treatise in its present state is perfectly adapted, in my opinion, to the wants of the medical student, who has not been liberally educated. The hurried courses now given at all the medical schools in America, allow but little time for natural philosophy. The student will find, that Webster has presented him with a concise and well digested exposition of every essential principle of the science.

AMOS EATON.

Troy, Jan. 17, 1824.

AUTHOR'S PREFACE.

The great object of science is to ameliorate the condition of man, by adding to those advantages which he naturally possesses. What would avail the deep and speculative inquiries by which the learned attempt to trace the source of Infinite Wisdom, if their labours did not produce benefit to their fellow creatures?

In this view the study of philosophy stands highly conspicuous, not only in the pleasure it affords in the pursuit, but in promoting our interests, supplying our necessities, and adding to the general happiness of mankind.

It is not a part of society, but the whole, that is interested in this kind of information; for well-grounded philosophy is the parent of arts, commerce and agriculture, which are the vital principles that promote the well-being of civilized states. Nor is it less efficacious in fixing the principles of religion: the more comprehensive our view of the divine productions in the creation of the universe, the more strong and lasting will be our conviction of the power, wisdom, and goodness of the Great Author of all things.

If, then, philosophical knowledge be of such essential advantage in the general pursuits of society, it surely becomes highly expedient to diffuse it in such a manner as to enable every class to obtain some portion of the whole.

As a number of learned works have been written on this subject, which are only calculated for persons of leisure and education, it cannot be an unworthy attempt to gather the fruits of these labours, and adapt them to more general use. It is, therefore, the humble endeavour of the author of this work, to collect and methodize those demonstrative truths, which have been drawn from the bosom of nature by the deep researches of the philosopher, and to render them plain and evident to those whose time and education will not enable them to draw their information from original sources.

By the exertions of able experimentalists, the love of physical science has been greatly extended; and the general class of society has become more interested in its pursuit, from a well-directed view of its utility. But even public lectures lose their effect upon a considerable part of the auditors, from the want of preparatory knowledge. If outlines of the subjects were previously fixed in the mind by an easy introductory work, the hearer would be prepared for the subject, and would gain much greater advantage from the lecture.

More than one half of the young people who are placed in public schools, are intended for those common avocations in life, which leave but a circumscribed portion of time to attain the various objects of education. It is, therefore, neither to be expected, nor is it intended, that they should acquire any thing more than a general knowledge of science. The first consideration then is, how to employ this small portion of time in such a manner as to produce the greatest advantage to the pupil. If it be admitted, that it is an object worthy of attention to instruct the youthful mind in physical knowledge, and to extend philosophy to the useful purposes of life, the subject will require such an arrangement that its acquisition may be rendered compatible with the time and ability of the pupil.

It is hoped that the following pages will not be found totally inadequate to this desirable purpose. Speculative theory and mathematical demonstrations have been as much avoided as possible, to make way for those useful and evident truths which are universally received ; but where demonstrations become indispensably necessary, they are introduced with as much brevity and perspicuity as the subject would admit.

In short, every endeavour has been exerted to make the work correspond with the intention ; which was to produce a cheap and comprehensive abridgment of Natural Philosophy, adapted to the understanding of the generality of persons ; and it is now left with a liberal public to determine on the utility and success of the undertaking.

JOHN WEBSTER.

ject worthy of attention to instruct the youth
d in physical knowledge, or to conduct them
to the ... of ... the superior ...
sure a ... as its acquisition may
ered compatible with the time and ability of
ll.

hoped that the following pages will not be
ntly inadequate ... to these purposes
tive theory and mathematical demonstration
en as much avoided as possible, to make way
e ... and ... experiments which are univer-
eresting; but where demonstrations became in-
ably necessary, they are introduced with as
ness and perspicuity as the subject would

every endeavour has been exerted to make
correspond with the intention; which was
uce a cheap and comprehensive abridgment of
l Philosophy, adapted to the understanding of
nerality of persons; and it is now left with the
public to determine as to the utility and success
undertaking.

JOHN WEBSTER.

CONTENTS.

GEOMETRICAL PRINCIPLES,

NECESSARY FOR READERS OF THIS WORK.

Fig. 1. *Right line*, or strait line, *a* *b* is drawn in the nearest direction between two points, as from *a* to *b*.

2. *Curved line*, is any line which is not strait.

3. *Circle*, is a plain figure, bounded by a continual line, called its periphery, every where equally distant from a point within it, called the centre.

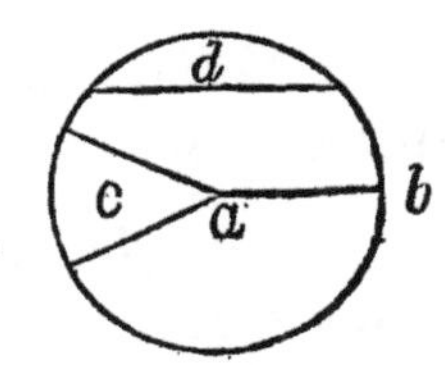

Radius, a line from the centre of a circle to its periphery, as *a b*.

Sector, a piece of a circle cut out by two radii, as *c*.

Segment, any piece of a circle less than half of it, which is cut off by one right or chord line, as *d*.

Arc, any piece of the periphery less than the whole.

4. *Elipse*, or oval, is a circular figure lengthened in one direction more than in the other. It may be produced by tying the two ends of a thread to two pins stuck in a board, the thread being longer than the distance between the pins ; and then moving a scratch pin around the fixed pins while it is held on all sides by the thread so bent as to form a loop for it.

Focuses, or foci, of the elipse, are the two points where the pins are fixed.

5. *Square,* a figure bounded by four equal sides, all meeting at right angles.

6. *Parallelogram,* same as a square, excepting that the opposite sides only are equal.

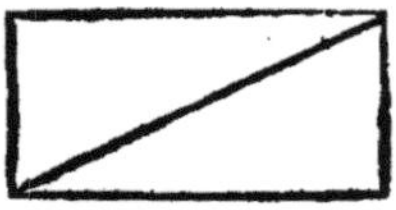

7. *Rhomb,* either of the two last figures, racked out of square.

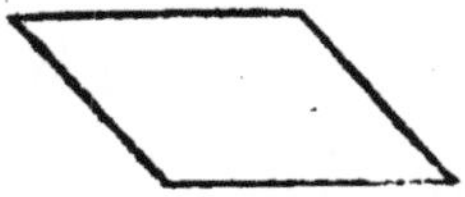

8. *Superficies,* having length and breadth, without any thickness.

9. *Solid,* having length, breadth and thickness.

10. *Angle,* a corner or point where two lines meet. If it is a square corner, it is called a *right angle.*

11. *Obtuse angle,* is wider open than a square.

12. *Acute angle,* not so wide open as a square.

13. *Triangle,* a figure bounded by three sides. If there is one right angle in it, it is a *right angled* triangle.

14. *Obtuse angled* triangle has one obtuse angle.

15. *Acute angled* triangle, has all three of the angles acute.

16. *Degrees of a circle,* are 360 ; each quarter or quadrant containing 90 degrees. The sine of an angle is a line let fall from one end of the arc of the angle, perpendicularly upon the opposite side. As the line *a c* is the sine of the angle *e.*

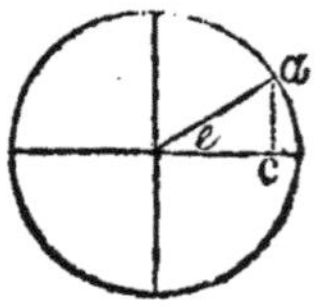

17. *Line of chords*, is a graduated line connecting the two ends of a graduated quadrant by a chord line. It is made by setting one foot of the dividers at *a*, and extending the other foot to each degree on the arc, and turning it down to the chord line. The chord line is used in plotting or projecting, when geometrical calculations are to be made.

18. *A triangle contains* 180 *degrees;* for a half circle contains 180 degrees, which is represented by the semi-circle *g h i*. By inspection, without taking the steps of a demonstration, the

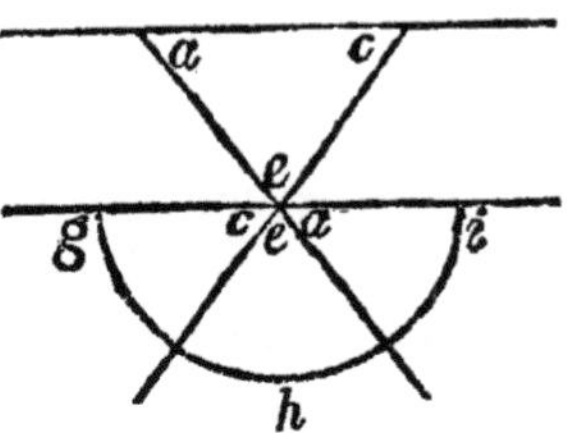

reader will perceive that the angles *a* and *a* are equal, also *c* and *c*, also *e* and *e*. Now as the angle *a c e* below the line *g i*, are measured by the semi-circle *g h i*, it follows that the angles *a* c *e*, in the triangle above the line *g i*, may be measured by the same; consequently, contains 180 degrees. The same elucidation may be given of all forms of the triangle.

19. *Geometrical trigonometry* gives all the sides and all the angles of a triangle, if two angles and one side, or two sides and one angle, are previously taken by the proper measures and observations.

From the following exemplification of this proposition, a reader of ordinary ingenuity, with no previous knowledge of trigonometry, may make all the applications which the following treatise requires. Draw the lines and angles given, then finish out the triangle in the only way in which it can be completed, without any random operation. The side *a b*, in the figure, and the angles at *a b* are given. Draw the given line 50 feet, calling any division of a scale a foot. Strike the arc *c d*, with a radius of 60 degrees, taken with the dividers from the line of chords on Gunter's scale. Take 70 degrees, being the angle at *a*, from the same line of chords, and set it off on the arc, which will extend to *d*. Draw the line from *a* through *d* indefinitely. Then with the sweep of 60, again strike the arc *e f*, and set off 80 degrees upon it, being the angle at *b*, which will extend to *e f*. Draw the line from *b* through *f* indefinitely, and it will cross the line which was drawn through *a* at *g*. Where these lines intersect each other, is the true place for the other angle. Measure the two new sides by the same scale by which the given line was laid down, and you have all the sides. Add the two given angles together, which will make 150 degrees. Subtract this sum from 180, the degrees always contained in a whole triangle, and the remainder will be 30, the degrees of the new angle at *g*.

Remark. Though all the important astronomical calculations are made by logarithmic trigonometry, still geometrical trigonometry, as here explained, will be sufficient for illustrating every calculation given in this work.

ELEMENTS

OF

Natural Philosophy.

NATURAL PHILOSOPHY *is that science which considers the powers of nature, the properties of natural bodies, and their action on each other.*

Illustration. Natural philosophy is distinguished from chemistry, by its treating on the laws of natural bodies in mass ; whereas chemistry treats of the laws of the constituent atoms of bodies, as those atoms bear relation to each other.

THREE LAWS OF MOTION.

First : *Every body continues in a state of rest, or moves uniformly in a right line, unless it be compelled to change that state by the action of some external force. This law of matter is called vis inertiae.*

Illustration. Thus a ball discharged from a cannon would persevere in its motion for ever, if it were not retarded by the resistance of the atmosphere and the operation of gravity. Or a top put in motion would have an endless revolution, if it were not impeded by the air and the friction produced by its point on the plane on which it moves. According to this law, the heavenly bodies also preserve their progressive motions undiminished in those regions which are void of all resistance.

Secondly: *The change of motion is always proportional to the moving force by which it is produced, and it is made in the line of direction in which that force is impressed.*

Illustration. By the first law, motion cannot be generated in a body, without some external impulse; and as the motion thus generated is directed in a right line with a velocity equal to the degree of impulse, the course of a body in motion can only be altered by a fresh impulse, and is then compounded of its own velocity and the impelling force; that is, the body will be either accelerated or retarded in a right-lined direction, in proportion to the compound force of the two impressions.

In the square A B C D, if a body be put in motion by an impelling force in the direction A B, and if at the same instant it be acted upon in an equal degree by another impulsion in the line A C, the body will be projected with a force and direction, compounded of the two impulses, which is represented by the diagonal line A D.

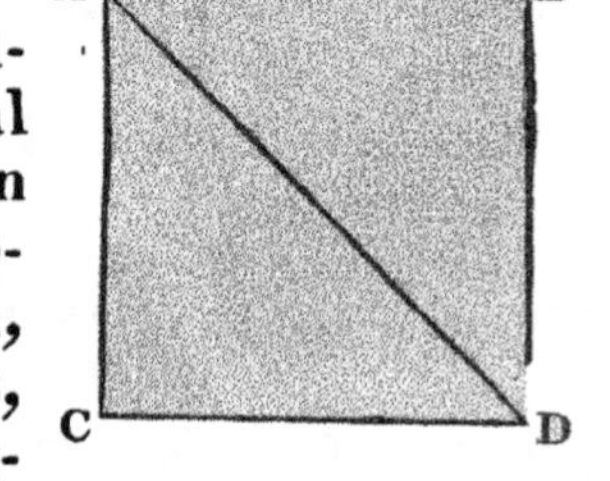

In like manner, if a ship at sea sail before the wind, in the line E F, due east, at the rate of eight miles an hour, and a current set from the north, in the direction E G, at the rate of four miles an hour; the vessel will be driven between the north and the east in the direction E H, compounded of the two acting forces, at the rate of nine miles an hour nearly.

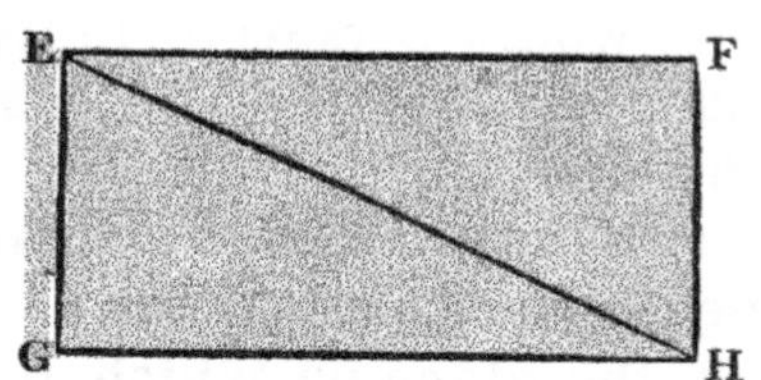

Thirdly: *Action and reaction are always equal and contrary. Or, the action of two bodies on each other is always equal, but in contrary directions.*

Illustration. Thus, if a stone is pressed by the hand, the reaction or compressive force of the stone on the hand, is equal to the pressure upon the stone. Or, when a horse draws a load, the power of the horse is diminished, or the animal is drawn back, with a force equal to that which puts the load in motion: for, if the weight of the load be increased till it is equal to the strength of the horse, it will remain at rest, although the whole force of the animal be in action. If a loadstone, and a piece of iron of equal weight, be suspended by strings near each other, the mutual force or attraction between them, will cause an equal action, and the two bodies will leave their respective positions with an equal impulse and velocity, and meet in a point equally distant from each.

If the bodies be unequal, they will meet in a point whose distance from the bodies will be reciprocally proportional to the difference of the powers.

If two floating vessels of equal magnitude be attached by a rope at some distance from each other, a force applied to the rope in either vessel, will mutually draw them together, with an equal velocity, till they meet in a point equidistant from their first position.

ATTRACTION.

Attraction is the cause, power, or principle, by which all bodies mutually tend towards each other.

Illustration. This universal principle is considered as one of the first agents of nature in all her operations. By this extraordinary power, the minutest particles of matter cohere, bodies are formed, and even the whole universe is governed by its influence; yet, after endless opinions, its cause is still concealed in the bosom of nature. We clearly view the effects of attraction, and decide on its laws, but human ingenuity has not been able to fathom its principle or essence.

Newton considers it as a power or virtue proceeding from bodies in every direction, which decreases in energy or effect, in proportion as the squares of the distance from the body increase; that is, at any given distance, it will be four times as great as at twice that distance, and nine times as great as at thrice the distance; and so on in like proportion.

This law, however, only relates to one branch of attraction, as it has different modifications in its different divisions: these are called attraction of *cohesion, electrical* attraction, *magnetical* attraction, and attraction of *gravitation.*

Attraction of Cohesion.

Attraction of cohesion is that force by which the particles of bodies mutually tend towards each other.

Illustration. It is the most powerful in the point of contact, or where the particles touch; at a little distance, it becomes considerably less; and when the particles are still farther removed, the effect is rendered insensible.

The power of corpuscular attraction, or the cohesion of particles of small bodies, may be shown by a variety of amusing experiments.

Take two leaden bullets, with a part cut away from each of their surfaces, so as to form a small plane, perfectly smooth and even. This being done, press the flat surfaces together, twisting the bullets with the fingers as they are pressed; then the parts which touch each will adhere or be attracted with such force, as to require a power of more than fifty pounds weight to separate them.

The twisting of the planes of the bullets serves only to bring the parts nearer together; for, as it is scarcely possible to cut the surfaces perfectly even, this twisting pressure tends, from the softness of the metal, to rub down the inequalities, to expel the air which is contained between the planes, and to bring a greater number of parts into contact.

As the formation of bodies arises from the adhesion or attraction of the particles of matter; if the metal in the above experiment were perfectly free from porosity, and the planes mathematically even, on joining them together, the parts in adhesion would be as firm and inseparable as any other parts of the bodies.

But as corpuscular attraction extends only to infinitely small distances, and as a considerable part of the surfaces cannot come into contact, not only from the porosity of the metal, but from the inequalities of the planes; the elasticity of the air, which is contained in the interstices, is perpetually endeavouring to force them asunder.

The planes can only adhere when the power of the parts in contact is greater than the natural gravity, and the elastic power of the air contained between them; therefore the cohesive force is proportionable to the number of parts that touch each other.

If oil, tallow, or any other unctuous body, be smeared on the surface of the planes, in such a manner as to exclude a principal portion of the air which is contained between them, the planes will adhere with much greater firmness ; so that plates of brass, silver or iron, of small dimensions, may be made to cohere with such force, as would require the united power of a number of men to pull them asunder. Experiment has shown, that plates not more than two inches in diameter, have taken a force of 950 lbs. weight to separate them, when the surfaces have been heated and smeared with boiling grease, and then left to cool before the power was applied. This adhesive power or quality in the particles of bodies, is not occasioned or aided by the gravitating weight of the atmosphere; for it is found by experiment, that it requires the same weight to separate them whether they be joined together in the open air or in vacuo.

By the attraction of adhesion the particles of a liquid arrange themselves in a spherical form.

Illustration. Rain, in falling from the clouds through the atmosphere, is formed into small spheres by the mutual attraction of the particles of water. The drops of rain which rest upon cabbage leaves, and other vegetables that are covered with a fine powder, also assume a spherical appearance from the same principle. Globules of quicksilver are formed in like manner, by the attraction of their parts, and incorporate by the same principle when different globules come into contact. If a piece of board, or any other plane be laid on the surface of water, it will require a power six times as great as the weight of the body to take it up perpendicularly.

These, and many other facts which daily occur in the common occupations of life, serve to show the universal tendency of that corpuscular attraction which exists between small bodies ; whilst the attraction of gravitation, extending to indefinite distances, causes all the regular changes and successions in the planetary system. Thus the Divine Being, by a different modification of the same incomprehensible principle, compounds and preserves the whole system of his works.

By that variety of the attraction of adhesion, called capillary attraction, liquids ascend between the contiguous surfaces of bodies.

Illustration. Take two plates of glass ground even, and place them edgewise, very near to each other, in a vessel of water, first wetting the insides of the plates; then the attracting power of the glass will raise the water which is contained between them considerably above the general surface of the fluid. This height is proportional to the distance of the plates from each other. If they be placed about the hundredth part of an inch asunder, the water will rise more than an inch above the common surface in the vessel.

The water ascends by the attraction of the plates, till the gravity or weight of the ascending fluid is equal to the power of attraction between the plates; and as the power of the glass by which the water is attracted is always the same, it is evident, that as the sides approach, the gravity of the fluid at equal heights becomes less, and consequently the elevation will be greater, before the weight of the water, or power of gravitation, becomes equal to the attractive force. If the plates be placed angularly, or touch each other at one of the ends, the water will rise in the form of a hyperbolic curve.

Capillary attraction is generally used to denote the ascent of fluids through small pipes or tubes, that compose a considerable part of the animal as well as vegetable body. In these tubes, which are as various in their number as they are different in capacity, by the energy of the living principle, nature conveys nutriment to supply the most distant branches of vegetation, where it could never arrive by the ordinary motion of fluids.

These tubes, as well as glass tubes, attract in an inverse proportion to their diameters, as the glass plates attract in proportion to their contiguity; that is, those tubes which are the smallest, raise the fluid to the greatest height, and the larger to a less height in a reciprocal proportion.

When the earth receives rain on its surface, the fluid is attracted through all the internal and contiguous parts; it is then absorbed by the roots of trees, plants, &c. and afterwards carried by capillary attraction to the most extended ramifications, through the multitudinous pores contained in the trunk and its branches.

Melted tallow and oil supply the flame of candles and lamps by the capillary attraction of the threads of the wick. Water poured round the bottom of a heap of sand, sugar, ashes, or any other porous substance, will diffuse itself till it has reached the summit. On this principle, perpetual springs are often supported on hills. This attracting power is likewise observable in lump sugar, sponge, linen, and many other bodies, when their lower extremities are dipped into water. In short, every porous or capillary substance is a conductor for the attracted fluid. This attracting power acts independently of atmospheric pressure; for if capillary tubes be placed in a vessel of water under an exhausted receiver, the fluid will ascend to the same height as when the experiment is made in the open air.

As a curious instance of the attraction of fluids through the pores of the skin, sailors, left without fresh water, frequently dip their clothes in the sea, and apply them wet to their bodies, which then attract the pure particles of the fluid, and by this mean they allay the extremity of thirst.

After cohesive and capillary attraction, the next division of this extraordinary power will lead us to take a slight view of electrical attraction; but as it will be necessary hereafter to enter into a more general detail of electricity, it will be sufficient, for the sake of order, to give here only some account of the leading principles of this subject.

Electrical Attraction.

Electrical attraction is that power which is excited by heat and friction, on the surfaces of glass, amber, and resinous bodies.

Illustration. If a cylindrical glass tube be warmed, and then rubbed briskly with an old silk handkerchief, and held near the downy part of a feather, or pieces of gold or brass leaf, some of them will immediately fly towards the tube.

If a glass globe have its axis placed horizontally, and some small flaxen threads be suspended from a semicircular wire, fastened over the upper part of its surface, when the globe is at rest, these threads will hang perpendicular and parallel to each other, according to their respective gravitations ; but on turning the globe briskly, its rotatory motion will communicate the same motion to the air that surrounds it, and this will impel or turn up the ends of the threads in the direction of the current of air which is produced by the motion of the globe ; but on applying a dry hand or rubber to the surface of the sphere, still continuing its motion, an electrical attraction will be excited, which will straighten the threads, draw them from their parallel position, and converge them in lines tending directly towards the centre of the globe.

The attraction of sealing wax, and other resinous substances, which is excited by rubbing them with a piece of woollen cloth, or on the coat sleeve, is too generally known to need any observation here. Amber has this peculiar quality of attracting light substances, which was the only instance of electrical attraction that was known for ages.

Magnetical Attraction.

Iron becomes magnetic by standing perpendicularly, or nearly so, on the earth ; the upper end becoming a north, and the lower end a south, magnetic pole.

Illustration. A bar of iron, a common poker, a

fire shovel or tongs, having been set nearly upright a short time, will attract an unmagnetized needle. If the upper end be brought in contact with one end of the needle, that end will become a south pole, and the other end a north pole; and if placed on a light chip on water, will arrange its poles accordingly. From this peculiar effect, it is supposed by some philosophers, that the earth admits a magnetical fluid, and that a part of it is retained in passing through the bar or metallic conductor.

Magnetic power acts in curved lines, forming circular connections between the north and south poles of the same magnet.

Illustration. The course of the magnetical effluvia may be made visible in the following manner:—Lay a sheet of white paper over a bar magnet, and sift some fine steel filings upon it; these, in falling upon the paper, will arrange themselves in the magnetical course, and form curved lines round both sides of the magnet, radiating from the two extremities, or the respective poles of the magnet.

A well-balanced bar of iron will dip at the north end in northern latitudes, and at the south end in southern latitudes, after being magnetized.

Illustration. Take a needle which is used for the compass, before it is magnetized, and balance it horizontally on a fine point; then take it off, and communicate the attraction either by a natural or artificial magnet, and place it again on the point: it will now lose its horizontal position, and one end of the needle will dip or sink downwards, making an angle with the surface of the earth. This is what is called the dip of the needle, and is supposed to arise from the subtile power that issues from the earth in an oblique direction, and which passes through the needle in its magnetical course.

Remarks. Natural magnets generally attract each other with less force than those which are made of steel. Rust and fire greatly injure, or totally destroy, the magnetical power.

Similar poles repel, and contrary poles attract, each other; and each pole communicates magnetism contrary to itself.

Illustration. Touch the point of a sewing needle with the north pole of a magnet; put it on a small chip upon water, and it will point south. Prepare another needle in the same manner, and their points will repel each other, but heads and points will attract.

Remarks. When magnets of different sizes are opposed to each other, the power of the superior magnet is capable of changing the poles of the inferior, which then causes them to attract one another.

To show the attractive power of a good magnet, let it be suspended from one end of a scale beam, and counterpoised by weights at the other; then fix a piece of flat iron about $\frac{2}{5}$ of an inch from the bottom of the magnet, and it will then descend and adhere to the iron. If they be again separated, and $4\frac{2}{5}$ grains be added to the opposite end of the beam, the weight will exactly balance the attractive power of the magnet, and oppose its descent; but if any part of the weight be taken away, the magnet will preponderate and descend as before. If it be placed at half the above distance, it will take four times the former weight, or about $17\frac{1}{2}$ grains to balance the scale beam; consequently the attractive force of the magnet, at the single distance from the iron, is to its force at double the distance, as 4 to 1; that is, reciprocally as the squares of the distances.

The power of the smaller magnets is generally greater in proportion to their weight, than that of the larger; for large magnets will seldom take up more than three or four times their weight, but the smaller will suspend ten or twelve, and in some instances twenty-four times their own weight.

Magnetic power may be strongly communicated to duly tempered steel, by friction and pressure, in contact with an artificial or natural magnet.

Illustration. Among the various modes that are recommended by different persons, the following appears the most simple, and yet sufficiently powerful for any common purpose.

Place two magnets, A and B, in a right line; laying the north end of one towards the south end of the other: (by the north or south end of the magnet, is meant, that end which would point towards the north or south pole, if the bar were balanced on a point; the north end is generally marked by a fine line cut across near the extremity of the bar.) After the magnets are thus placed, let the bar C, which is to receive the magnetic power, and consists of steel tempered but little higher than a common knife-blade, rest upon the two extremities of the former, placing the end that is intended for the north upon the south end of the magnet that supports it; this being done, take two other magnetical bars, D and E, and bring the north and south ends of them together upon the middle of the bar C, raising up the opposite ends till the angles which are formed on the bar become equal; then separate E and D, by drawing them different ways to each end of C, still keeping the extremes of D E at equal distances from the plane; after they are drawn off, join them together about four inches above the plane C, and again place them in contact on the middle of the bar; after this has been repeated four or five times, turn each of the other three sides upward, and pursue the same operation a dozen times or more, which will communicate a strong and permanent magnetism to the bar of steel.

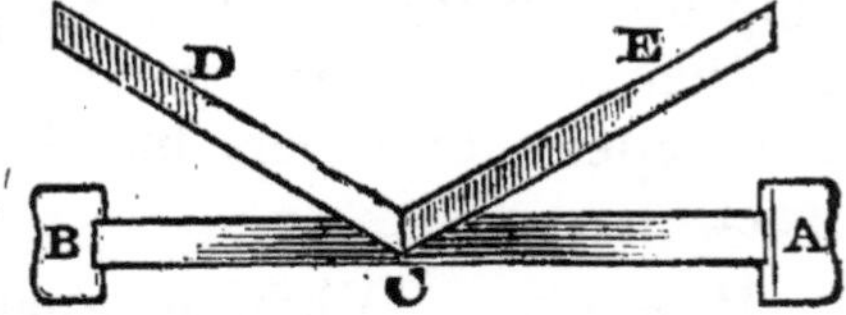

The magnetic power may be easily communicated, by what is called the horse-shoe magnet, from its resemblance to that form. Lay the bar to be magnetized upon the extremes of two magnets, as in the preceding directions, and place the two ends of the horse-shoe magnet upon the middle of the bar, observing that the north pole of the magnet is placed towards that which is intended for the south pole in the bar; then draw it backwards and forwards five or six times, and afterwards take off the magnet from the middle of the bar. The same mode must be followed on each of the three other sides.

Attraction of Gravitation.

Gravitation is that force with which bodies tend towards the centre of the earth, or by which they fall perpendicularly to its surface.

Illustration. Gravitation seems to differ from corpuscular attraction, only as a part differs from a whole: the attractive power which singly unites the particles of smaller bodies, may form that gravitating power, in the aggregate, which governs the system of the universe; thus, considering the attractive influence of bodies as proportional to their magnitudes, the less will be governed by the greater, and those which are on, or near, the surface of the earth, will tend towards its centre: Bodies not only gravitate towards the earth, but likewise towards great elevations or mountains on different parts of its surface; for if a ball be suspended by a line, and placed on different sides of a high mountain, it will gravitate on every side towards the mountain.

The power of gravity gives the same velocity to all bodies; therefore, taking away the resistance of the air, or the medium through which they fall, the descent of all bodies from the same height will be performed in the same time, whether they be great or small, light or heavy.

Illustration. If a piece of gold and a feather, or any other bodies, the specific gravities of which are different, be dropped from a given height, through the atmosphere, the superior gravity of the gold will more effectually overcome the resistance of the air, than the inferior weight of the feather, and consequently it will fall much sooner to the ground; but if they both fall at the same instant, from the slip or dropper of an exhausted receiver, they will arrive at the bottom in equal times; for as the resisting medium of the air is here taken away, the bodies descend with equal velocities.

Falling bodies gravitate with an increasing velocity as they approach the surface of the earth.

Illustration. This accelerated motion is produced by the constant power of gravity, which, by adding a fresh impulse at every instant, gives an additional velocity and an increasing motion in every moment of time. The space through which a body falls by the power of gravity in the latitude of London, is $16\frac{1}{12}$ feet, in the first second of time, four times that distance in two seconds, eight times in three seconds, and sixteen times in four seconds; increasing in velocity according to the squares of the distance through which the body descends.

Let A B represent the equal parts of time through which a body is accelerated in falling; then the velocity acquired in passing through each space by the continual impulse of gravity, is relatively as the lengths of the parallels *a* 1, *b* 2, *c* 3, &c., throughout the whole figure; that is, the velocity increases as the lines increase in length, and the quantity of the velocity* is equal to the square of the time; for if 5 *e* represent the velocity acquired in falling through the space A 5, the sum of the velocities,* or all

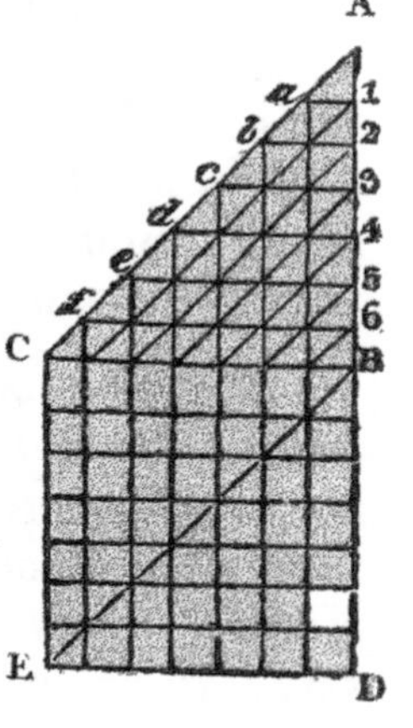

* That is, the whole space descended through, will be proportional to the square of the time of descent. *Patterson.*

the similar triangles taken together, which are contained in A 5 *e*, will amount to 25, which is equal to the square of the time A 5; and so on of any other point of time.

The spaces described by a uniform motion with the last acquired velocity, during a time equal to that of the acceleration from the beginning, will be double the space described by the accelerated motion.

Illustration. If B D be equal in time to A B, and B C and its parallels express the equable time of motion, the parallels likewise express the velocities, as in the preceding case; but they are double the spaces of the accelerated motion.

The law of acceleration in bodies descending perpendicularly, holds equally in point of time with those bodies that are projected.

Illustration. If a stone be dropped from the top of a tower, and another of the same weight be thrown horizontally at the same instant, the two different bodies will reach the ground at the same moment of time. Even if a ball were fired with any force horizontally from a cannon, on the top of a tower, it would describe the line of its course in the same time that another ball would fall perpendicularly from the top to the bottom, supposing them to meet with no resistance from the air.

For if A D be the horizontal line of direction, which the ball would move in by the force of the gunpowder, and A B the perpendicular line of gravitation, then, according to the second law of motion, the ball will not pass in either direction, but in the line A C, compounded of them both: again, by the same law, it would pass through the curve A *a*, in the same

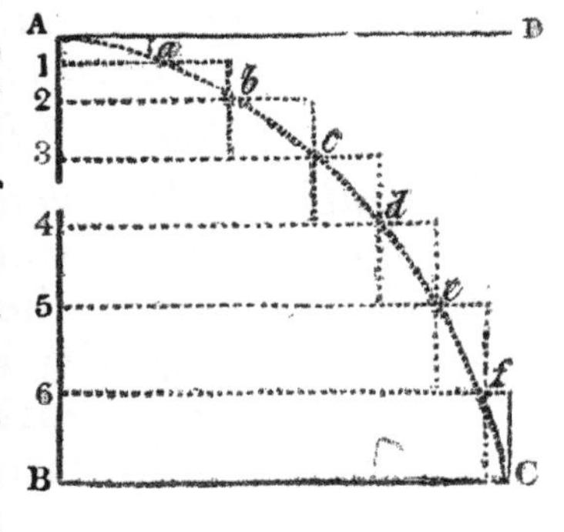

time that its gravity alone would carry it from A to A 1; therefore the time of describing A *a*, is equal to the time of describing A 1 : A *b*, to A 2 : A *c*, to A 3; and so on for the whole, making the sum of the times of the direction of the ball A C, equal to the times of the gravity A B.

If a ball be dropped from the topmast of a ship, even supposing the vessel to sail through the water with a velocity of ten miles an hour, it will fall exactly at the bottom of the mast.

Illustration. Let A C be the ship's mast, and its position when the ball is dropped from A; C D the distance sailed during the time of the descent; and B D the second situation, when the ball strikes the deck. Then the force, or projecting velocity of the vessel, C D=A B carries it towards B, and the force of gravity acts from A towards C; but the compound force carries the ball to D, in the direction A D, and in the same time that it would fall from A to C, if the vessel were at rest. And as the velocity of the ball and vessel are equal, it apparently drops perpendicularly down the side of the mast, as it describes the curve of projection; but a person in a vessel at anchor, at some distance, would observe its curvilinear direction.

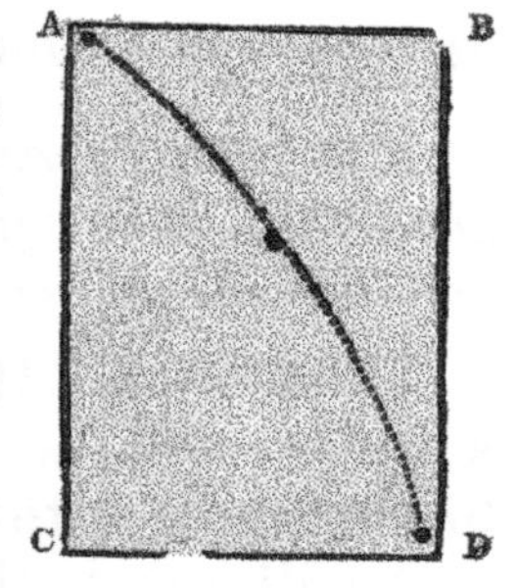

The power of gravitation is greatest at the surface of the earth, and decreases both upwards and downwards, but in a different proportion.

Illustration. In ascending, the gravity decreases as the squares of the distance from the centre increase; for at the distance of the earth's semidiameter from its surface, the gravity is not more than a fourth of that power on its surface. The force of gravity downwards from the earth's surface, is in a direct ratio as the distance from the centre; for at half the semi-diameter from its centre, the power of gravity is only equal to half the power at the surface; at a quarter, one fourth; and so on for any given distance in like proportion.

The power of gravity retards bodies that are thrown upwards in the same proportion that it accelerates those that fall; so that the times of ascent and descent are equal to and from the same height.

Illustration. If a stone be thrown from D towards B, and B C be the perpendicular line of gravitation; the stone will be retarded in its ascent, or the gravity B C will overcome the impetus, in proportion to the decreasing parallels, *a* 10; *b* 9; *c* 8, &c. till the whole projecting force is destroyed. When the stone returns, it will fall from B towards C, with an accelerating force, increasing as the parallels increase, till it reaches the surface at D: so that according to the laws of gravity, the ascending and descending times are equal.

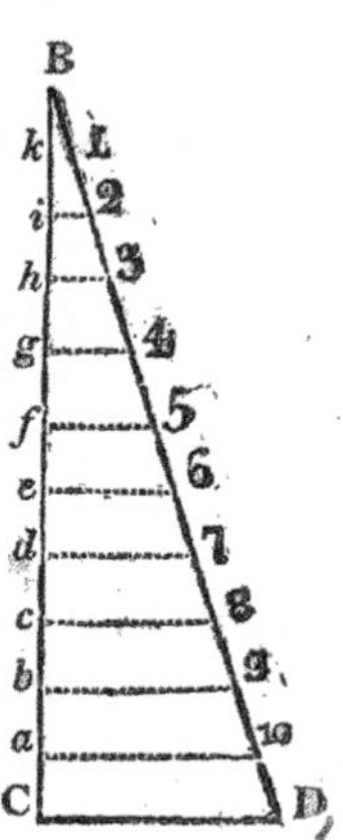

COLLISION.

Remarks. The collision of these bodies chiefly depends on the third law of motion, in which the action and reaction are equal and contrary. For if two bodies strike each other, their motions are equally affected, and their collision produces equal changes, but in contrary directions.

Bodies which are devoid of elasticity will not separate after the stroke, but will move on with half the velocity and all the momentum that the striking body had acquired before the impact, supposing the bodies to be equal.

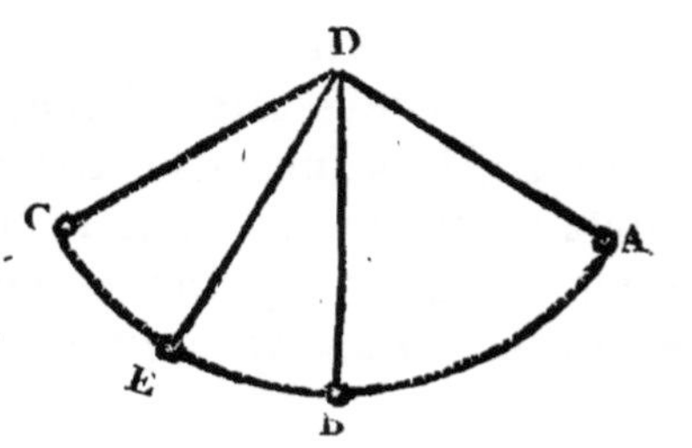

Illustration. Let two balls A B, made of clay or any other non-elastic body, be suspended in such a manner that their surfaces may just touch each other when they are at rest; then let A fall from any given height, and it will acquire such a velocity in describing the arc A B, as would carry it to an equal height C, if it were not obstructed by the other ball B; but on striking B, which is of equal magnitude, it communicates one half of its momentum, and the two balls move on together to E$=\frac{1}{2}$ B C. In experiments of this kind, whether with elastic or non-elastic bodies, the theory will vary in some degree from the practice, not only from the imperfection of the bodies, but from the resistance of the medium through which they fall.

Remarks. What is meant by the collision of elastic bodies, is when the particles give way at the point of impact, but restore themselves to their first position after the pressure is removed. The most perfect elastic bodies are ivory, glass, hardened steel. and some compound metals.

To show the elasticity of these bodies: If two balls of the above description be suspended from a common centre, and one of them has its surface thinly painted; then as it hangs in contact with the other, it will make a small mark on the side; but if the painted ball be raised to a certain height, and then suffered to fall against the other, the second impression on the receiving ball will be considerably larger than the first, which could not have happened if the elasticity of the balls had not suffered them to be indented by collision.

When a body, which is perfectly elastic, strikes another of the same kind and magnitude, the striking body will communicate the whole of its momentum to the other, and afterwards remain at rest.

Illustration. Let two balls of ivory, C B, be suspended from A; on suffering B to fall freely on C, it will lose the whole of the motion which it had acquired in its descent thro' the arc B C, and C will be driven up to D, in the same manner as B would have been carried to that point by its acquired velocity, if it had not been obstructed by C.

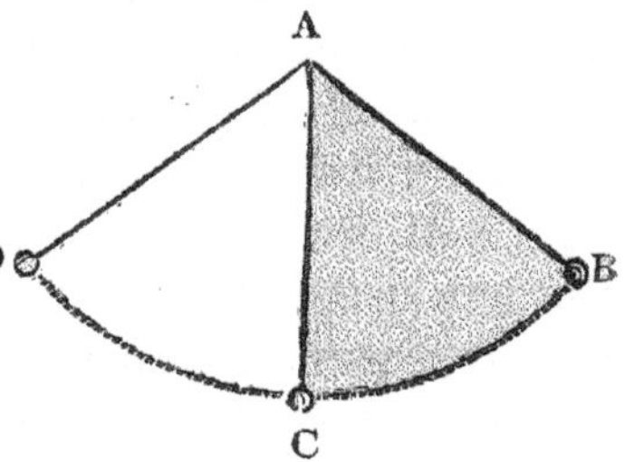

If four, or any other number of equal elastic balls, be suspended on centres near each other, and D be let fall from any given height; then as it strikes C, it will communicate the whole of its motion to it; and this will pass through the centres of C and B to A, leaving B C D quiescent, whilst A is driven off with a velocity equal to that which would have been communicated by D, if it had not been obstructed by the intervening bodies B C.

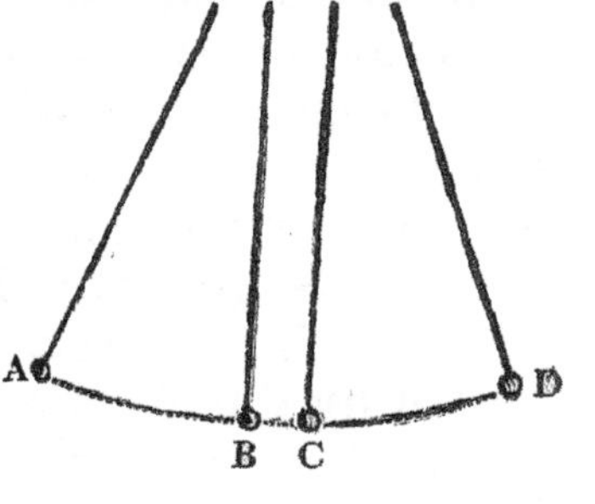

Whatever number of balls may be let fall on one side, the same number will have an equal motion on the other, and the intermediate balls will remain motionless.

PENDULUMS.

A pendulum, when put in motion, will descend through one half of an arc by its own gravity, and ascend the other half by the velocity which it has acquired in its descent.

Illustration. If A be a weight suspended by a line or wire, from the centre or point of suspension C; when it is let fall from D, it will descend through the arc D A, by its own gravity, and acquire such a velocity as it would obtain in falling from E to A.

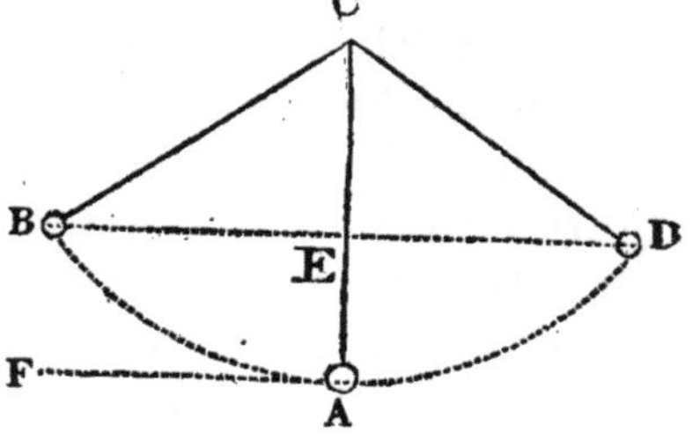

By the first law of motion, the weight would fly off in a tangent, or straight line A F, if it were not retained by the string; which, with the velocity the weight has acquired at A, conducts it to B, and the whole arc D B forms one oscillation. At B, which is in the same horizontal line with the opposite point D, it loses its motion, and then returns back again by the force of its gravity: thus the pendulum would continue moving forever, if it were not perpetually retarded by the friction of the cord upon the point of suspension, and the resisting medium of the air through which it vibrates; but these gradually retard its motion, so that each oscillation becomes something less than the preceding one, till at length the ball rests in the line of suspension C A.

The time of the vibration of pendulums is as the square root of their lengths.

Illustration. If one pendulum be four times longer than another, it will vibrate half as fast; and if it be nine times longer, one-third as fast; if sixteen times, one-fourth as fast; and so on according to the square root of the length, as before stated.

The centrifugal force, or rotatory motion of the earth round its axis, which causes it to swell out at the equator, and flattens it towards its poles, likewise causes pendulums to vibrate in unequal times in different latitudes.

Illustration. As the repulsive or centrifugal force is greatest at the equator, and lessens gradually towards the poles, the vibration of the same pendulum will become slower by the increasing resistance and diameter, as it is taken towards the equator, and faster by the decreasing power and diameter, as it approaches the poles of the earth.

Remarks. A pendulum of 39.2 inches in length, vibrates seconds in the latitude of London; and as a second forms an aliquot part of the time occupied in a diurnal revolution of the earth, the utility of the pendulum is sufficiently obvious, in marking the different divisions of that period of time.

All metalline bodies expand and contract by heat and cold, which prevents the common wire pendulum from being always quite correct in its length, even in the same latitude; this is partly remedied by forming the rod of bars of different metals, called the gridiron pendulum, the various actions of which counteract each other, and render it more permanently accurate in its dimensions. A uniform rod, one-third longer than a pendulum which is formed by wire and a weight, will vibrate in the same time, and the centre of oscillation in the weight will be the centre of percussion in the rod. By the centre of percussion, is meant that part of a rod or stick which produces the greatest impression in striking a blow.

CENTRE OF GRAVITY.

The centre of gravity is that point about which all the parts of a body exactly balance one another; therefore, if this point be supported, the body will be at rest, whatever be its form; and as this point may be conceived as the concentration of the weight of the body, it is hence called the centre of gravity.

Illustration. If two bodies of equal weight be fastened to the extremities of a uniform rod, the point of suspension, or centre of gravity,

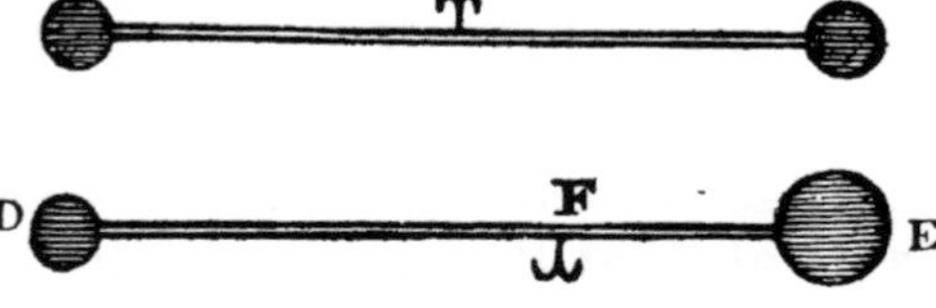

will be in the middle, or equidistant from the two extremities, in the manner of a scale-beam. But if the bodies be of unequal weights, the point of gravitation in the rod will be according to these weights; that is, if E be to D as 3 to 1, then F E will be to D F as 1 to 3, or the length of the arms from the point of suspension will be inversely proportional to the weights.

The centre of gravity in any plain figure may be found in the following manner: Let the body be freely suspended by a string, and apply a plumb-line to A, the point of suspension; this will pass over the centre of gravity somewhere in the line A B, which falls beneath the centre of suspension; then mark the line A B, and hold the plumb-line again from C, another point of suspension, and the point E, where the lines cross each other, will be the centre of gravity.

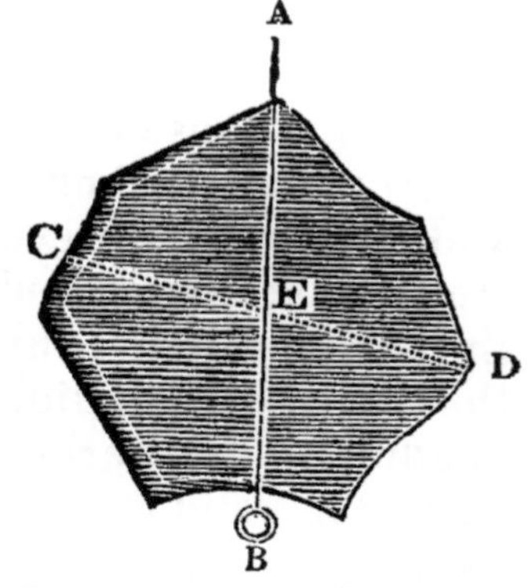

The centre of gravity in a square or flattened bar, is easily found by balancing it on the blunt edge of a knife. After fixing the knife,

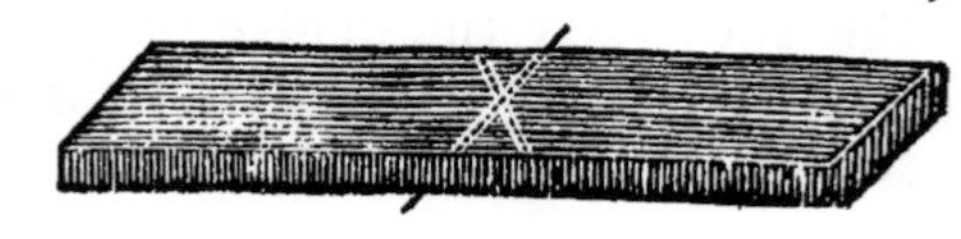

balance the bar with the ends a little inclining towards it, and then mark the line of gravitation; afterwards reverse the position and balance it again, and the point where the edge crosses the first line is called the point of suspension, or centre of gravity.

If a body be placed on a horizontal plane, and a perpendicular line from the centre of gravity fall within its base, it will stand; if the line fall beyond the base, the body cannot support itself.

Illustration. If a bar be placed endwise on the edge of a table, and the line of direction fall within the base D C, the centre of gravity will be supported by the table, and the body will not fall, although the top overhang its base; but if it be still more inclined, so that the line of direction A C comes beyond the point C of the base, the centre of gravity loses its support, and the body falls to the ground.

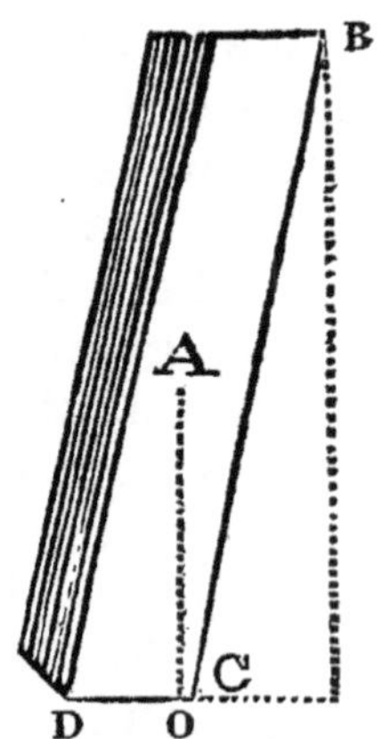

Inclining walls and old buildings support themselves from the above principle, and remain with their tops overhanging their bases for a number of years.

The tower of Pisa, in Italy, amongst many others, is a remarkable instance in proof of this law of gravitation; for the top of the tower overhangs its base sixteen feet, so that strangers pass by it with terror, lest it should fall on their heads; but as the line of direction falls within the base of the building, it still remains supported; and as it has stood some centuries in this state, it is probable that it may stand many more, if the cement by which it is held together should not perish.

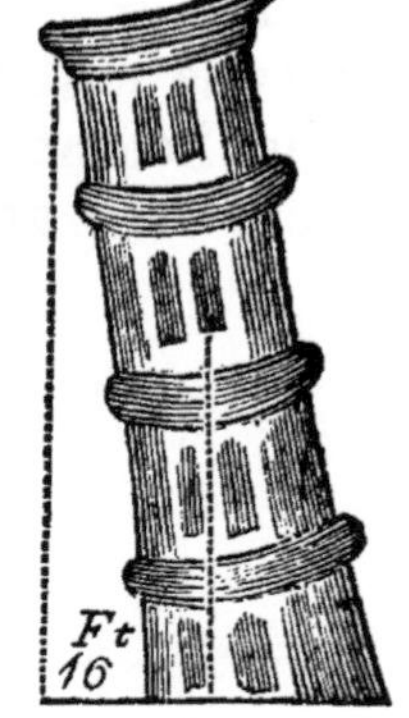

A body stands the most firmly when the base is

broad, and the line of direction falls in the centre; consequently the more narrow the base, and the more the line of direction approaches its extremity, the greater is its danger of falling.

If a piece of board be placed on the edge of a table, and the line of gravity fall beyond it, it will of course drop to the ground; but if two wires, loaded at one end, have the other fastened to the upper side of the board, and the weights rest over the table, the line of direction will fall nearer the weights, and the board will remain supported by the edge of the table.

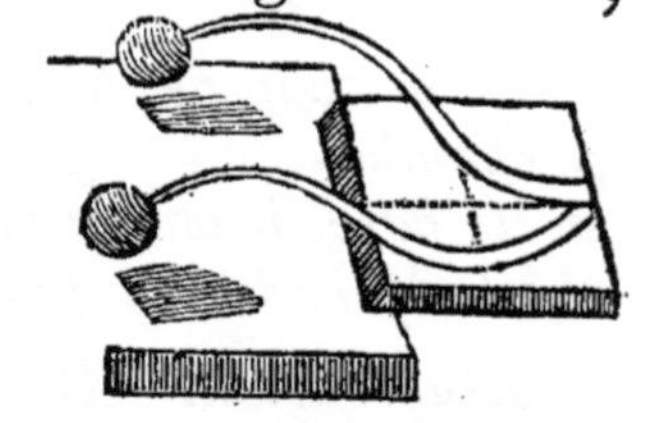

If A be a double cone, and the radius of its base C A, be greater than the height B D, of the inclined plane B C and C E; on placing it towards the angular point, between the sides of the plane, the cone will ascend, and the centre of gravity seem to move upwards; but the reverse is the fact; for the centre of gravity descends with the cone as it rolls towards its extremities, between the diverging sides of the wire. If the height of the plane B D, be equal to the radius, or half the thickest part of the double cone, the cone will remain at rest on any part of the frame. If the height be made greater than the radius, it will descend towards the angular point.

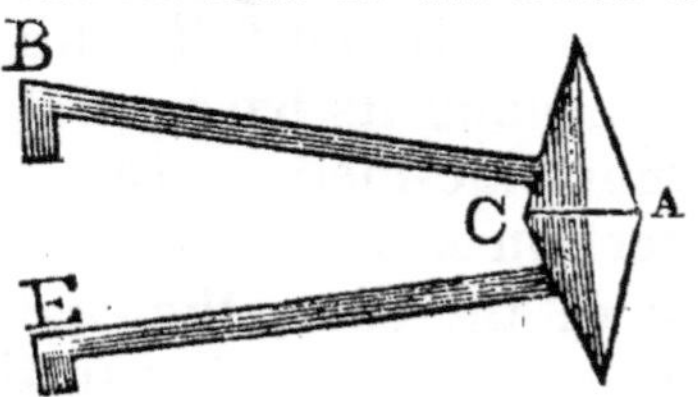

MECHANIC POWERS.

These powers are simple instruments or machines in the hands of man, by which he is enabled either to expedite motion, or to overcome such resistances as his natural strength could not effect without them.

The simple mechanic powers are exemplified in the *Lever*, *Wheel* and *Axle*, *Pulley*, *Inclined Plane*, *Wedge*, and *Screw*.

Remarks. Compound machines are formed from these simple powers; even the most complex machine is, in a mechanical sense, nothing more than a combination of the above simple forces.

In considering their operations theoretically, the principles are mathematically just; but in a practical application, some allowances must be made for weight, thickness, and friction.

The Lever,

Is divided into the prying lever, the lifting lever, and the radial lever.

The prying lever has the weight on the opposite side of the fulcrum from the power applied, and is in equilibrio when the long end is to the short as the weight is to the power.

Illustration. The operation of a lever of this description, may be represented by a common poker in the act of stirring the fire. Here, the poker is the lever; the bar of the grate, the bearing place or fulcrum; and the fuel contained in it, the weight to be raised.

The prying lever A B, has the fulcrum or prop C, between the weight and the power; the shorter arm A C, is made thick and heavy, that it may ba-

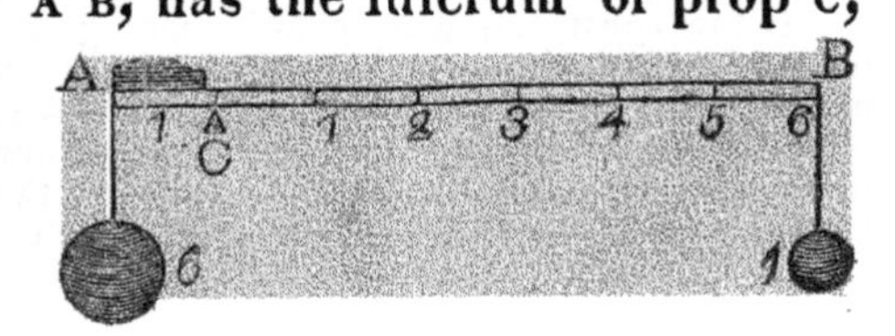

lance the longer end C B, so that the lever may be considered without weight, as it turns upon the fulcrum C. The advantage or additional power acquired by the use of the lever, is in proportion to the difference of the arms; for if the arm C B, be divided into six parts, and the other arm A C, be equal to one of these parts, a weight of six pounds suspended from A will be balanced by hanging one pound at B, the opposite end. Here the six pounds may be taken as the resistance, and the opposite pound as the power; so that the resistance is as much nearer to the prop, as the weight or power is lighter than the resistance.

Suppose a stone of a ton weight, to be fastened by a rope to the end of a lever $10\frac{1}{2}$ feet long, and the prop to be placed under it, at the distance of six inches from the end; the power of a hundred weight at the other extremity will raise the weight from the ground. Thus, by increasing the longer arm, even the ordinary power of a man may be made equal to the greatest resistances.

A balanced scale-beam is a lever of this kind, with its prop placed in the middle, and the centre of gravity directly under it.*

If two bodies, F and G, of equal weight, be suspended at equal distances from the prop, the centre of gravity will still rest beneath the point of suspension; for as the weights and distances are equal, the powers are equal; therefore, neither end can preponderate. But if F be moved to K, the centre of gravity E, moves towards C, and falls beyond D, the point of suspension; in which case it becomes unsupported, and

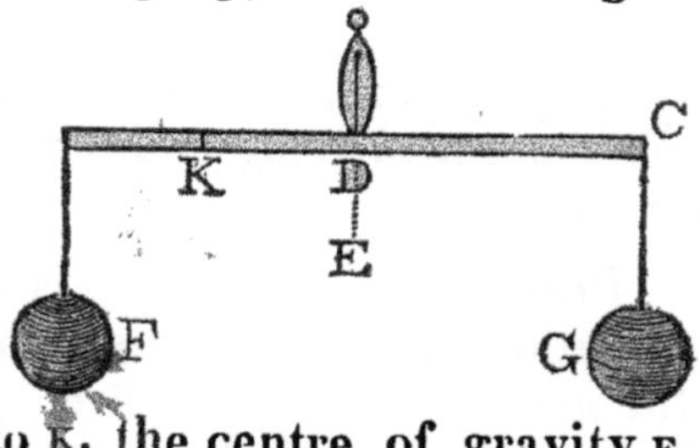

* In the construction of scale-beams, great care must be taken to place the centre of motion and the two centres of suspension, (at the extremities of the beam) in the same right line; otherwise, when the beam is moved out of its horizontal position, one end will approach nearer to the vertical line passing through the centre of motion than the other, and the lesser weight might actually appear to preponderate. *Patterson.*

the end of the beam C, descends towards the line D E; for by an invariable law of nature, the centre of gravity is impelled towards the line of suspension. This likewise shows that a smaller weight than G will balance the opposite weight F, when it is suspended from K. The steelyard acts on this principle, for if the longer end D C, were equal to six times the shorter K D, then a body suspended from C, would balance six times its weight, suspended from K.

Scissars, snuffers, pincers, &c. are made up of two levers, and the fulcrum is the rivet that fastens them together.

The lifting lever has the weight on the same side of the fulcrum with the power applied, and is in equilibrio when the long end is to the short as the weight is to the power.

Illustration. If one end of the lever A B C, be supported at A, and A C be seven times as long as A B; seven pounds suspended at D, will be balanced by one pound at G, the opposite end: thus, the common power of a man would be increased seven fold in raising any great weight suspended from B; but to raise the weight a foot, or an inch, from B to E, the end C must pass through seven times that space, or from C to F; so that, what is gained in power is lost in time, in this, as well as in every other augmentation of force by mechanic power.

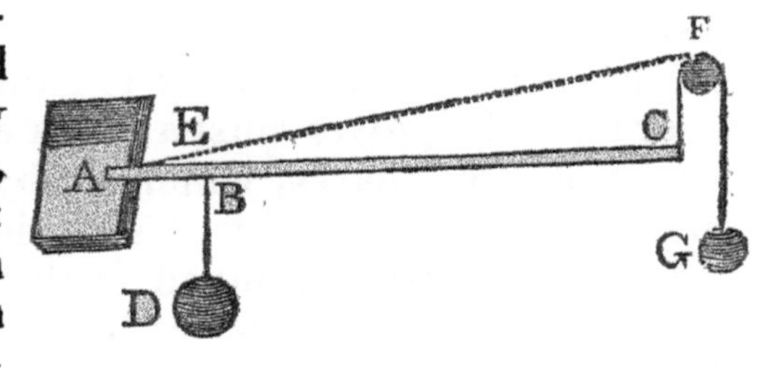

A wheel-barrow may be considered as a lever of this kind; A being the place of the wheel, C the handles, and B the load in the barrow; under this augmentation of power, a much greater weight may be moved by means of this machine, than could be supported on the shoulders of a single person.

A patten-maker's knife acts as a lever of the same

kind ; for the end which is fastened, is the prop, the power is at the handle, and the wood which is cut, the resistance. Here the power is increased as the resistance approaches the prop.

The radial lever has the power applied between the weight and the fulcrum ; and is in equilibrio when the long end is to the short, as the power is to the weight.

Illustration. If the lever A B be divided into seven equal parts, with the one end A, fixed under a table, and a ball of one pound be suspended from the opposite end, it will require a force of seven pounds, pulling upwards from the first division C, to balance the opposite power. Here the smaller weight, in rising or falling, has seven times the velocity, and describes seven times the space of the larger.

No mechanical advantage is gained by this kind of lever ; for the power must always exceed the weight. A pair of wool-shares, which act by a pressure in the middle, is a lever of this description. But this kind of lever is mostly employed in the action of the animal body ; for the bones are levers, the joints of them the fulcra, and the muscles the power that gives motion to the whole. It is always employed to expedite motion at the expense of power.

Wheel and Axis.

The wheel and axis is a perpetual lever, having its fulcrum in the centre of the axis and the wheel ; the longer part of the lever being the radius of the wheel, and the shorter part being the radius of the axis.

Illustration. In the figure, let A be the axis, the

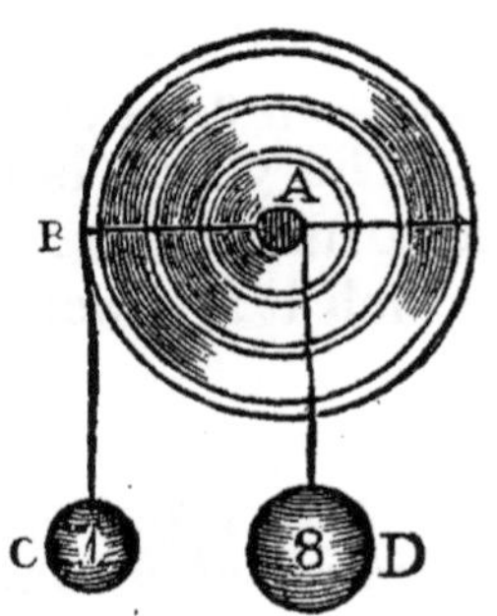

radius of which is six inches, and A B, 48 inches, the semidiameter of the wheel: then a weight of eight pounds suspended from A, will be counterbalanced by one pound hanging from B; for as the acting part of the lever, or radius of the wheel, is eight times as long as the resisting part, or radius of the axle, the power C will balance eight times its weight, suspended from A, which accords with the principle of the lever. The advantage gained by the power, is lost in the time, as in the preceding examples; for whilst D ascends through the space of a foot, C must descend through eight times that distance.

There are many modes of applying this mechanical power; as, by a handle which is used to wind up a jack, or raise a bucket of water in a well; here the advantage is in proportion to the thinness of the roller and length of the crank, or the distance of the hand or power from the axis. The capstan is another machine formed on this principle; the upright post which turns on a spindle, and receives the coil of the rope, may be considered the axis of the wheel, and the levers which are fixed into the post of the radii, at the extremities of which the power is applied. The power which is gained by this machine, increases as the difference increases between the radius of the upright post and the length of the levers which turn it. As the coils of rope increase upon the post, the radius of the axle becomes larger, and therefore lessens the force of the lever. Cranes of various forms for loading and unloading goods, as well as the capstans, are generally constructed according to the principle of the axis and wheel.

The Pulley.

Remarks. This mechanical power is formed by a small wheel, made of wood or metal, with a groove in its circumference, which is placed in a frame, and turns on an axis.

The wheel is called the sheave; the axis on which it turns, is the gudgeon or pin; and the frame in which it is placed, is called the block.

A single sheave fixed by its pin, is a perpetual prying lever, of the scale-beam variety, and produces no mechanical advantage.

Illustration. Let the pulley A support two equal weights P and W, from the ends of a cord which passes in a groove over the sheave; then P W will counterpoise each other, and the bodies will be at rest, in the same manner as if the cord was cut into two parts, and each tied to the end of a balanced beam; which is represented by the dotted line D A E.

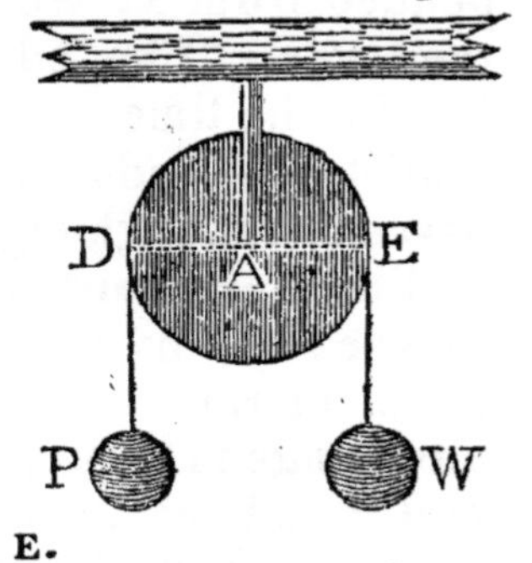

Therefore, supposing P to be the power, and W the weight, a single pulley does not increase or diminish the mechanic effect.

The principal advantage which arises from a single pulley, is in the mode by which the power is applied; for in raising a loaded basket or bucket, by a line passed over a pulley, the force acts downwards; which is much less laborious than when it is applied in an opposite direction.

A moveable sheave with one end of its rope fixed above, is a perpetual lifting lever, and always gives a double advantage to the power applied, over the weight.

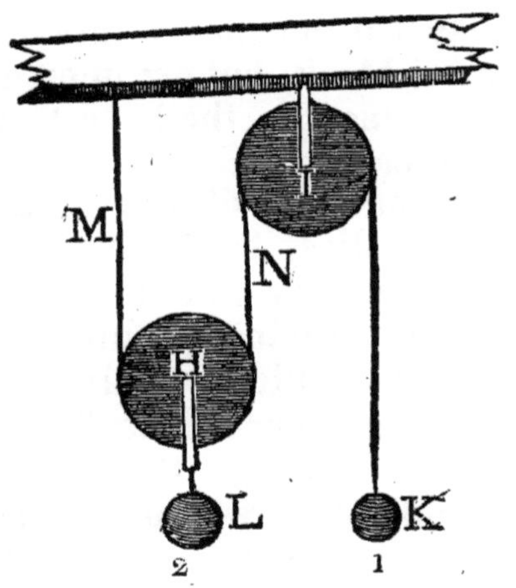

Illustration. When one end of the tackle or cord is fastened to a hook in the beam, and the other passes through the groove of the moving sheave H, and from this over another fixed at I, the power K will support double its own weight suspended at L; for it is obvious that the cords M and N each support one half of the weight; therefore, half the weight of L, which is supported by N, is again counterpoised by K; this being equal to half the weight supported by N, it is likewise equal to half the weight of L, and the power gained is in the proportion of two to one. Here, again, the advantage which is gained by the power is lost in the time, as the power K moves through twice the space of the weight L; for if the pulley H rise a foot, the cords M and N must be shortened a foot each, which gives two feet to K, the descending power. This difference of motion increases according to the number of sheaves in the block; for if there were three sheaves in each block, there would be six cords, and the power would descend six times as fast as the weight would rise.

If a fixed and a moving block have each of them three sheaves, and a weight be suspended from the lower block, it will be equipoised by a power equal to one sixth of the weight; for as the weight is supported by six lines from the three sheaves in each block, the weight is equally divided amongst them, in the same manner as in the preceding example; so that the power obtained is as six to one, or a weight of a hundred pounds at P, will balance six hundred at W; but P must move through six times the space of W.

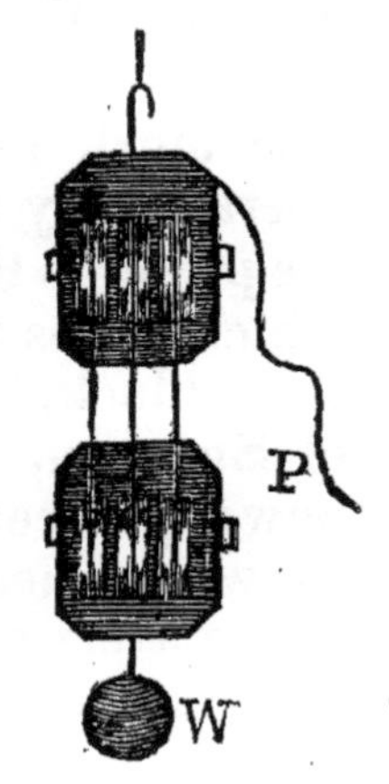

When the sheaves are not fastened together in the lower block, but act upon each other, and the weight is fastened to the lowest, the power will be greatly increased.

If one end of the cords which pass through the four pullies, be fastened to a beam, and the other to the block of the adjoining pulley, the weight would be divided in such a manner amongst the different pullies, that sixteen pounds at W, would be balanced by two pounds at P; for if the cord *g c* suspend sixteen pounds from the block W, according to what has been said, each part of the line supports one half of the weight, and the half which is supported by *c* is again divided into two equal parts by *f* and *b*, and *b* sustains one half of *c*, or a fourth of the whole weight W; the weight at *b* is again divided and balanced by P the power: so that the power is equal to the whole weight at W, a quarter at *c*, and one eighth at *b*. The uppermost pulley gives no increase of power, and it is placed merely for the convenience of pulling downwards from P, rather than upwards at *a*.

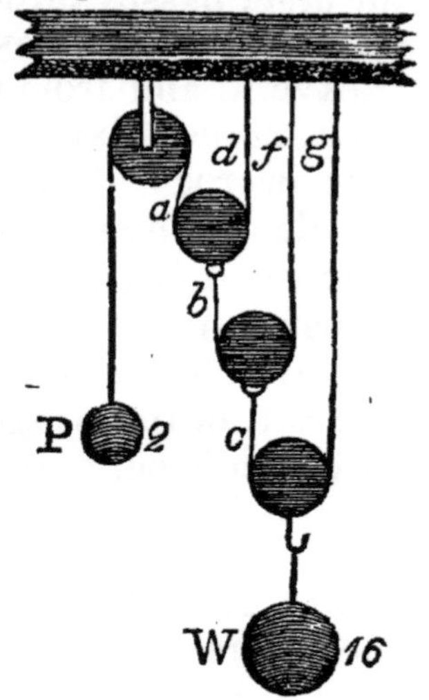

There are many other combinations of blocks, to diminish the weight and give increase to the power; but a number of sheaves in the same block not only lose much in the time, but are obstructed in their operation by the cheeks of the block, from applying the power to the outside sheave. This inconvenience, which arises from having a number of sheaves in the same block, has been greatly lessened by an invention of Smeaton, who makes the rope that sustains the power proceed from the middle sheave in the block, by which means the power acts upon all the sheaves in the same parallel direction.

Inclined Plane.

The power applied for raising a weight upon an inclined plane, must be to the weight as the height of the elevated end of the plane is to its length.

Illustration. If the perpendicular height B C be equal to one third of the base A C, and if the line which passes over the pulley E be fastened to the axis of a cylinder of three pounds weight at D; one pound suspended at the opposite end of the cord F will equipoise the three pounds at D, and the two powers will be at rest.

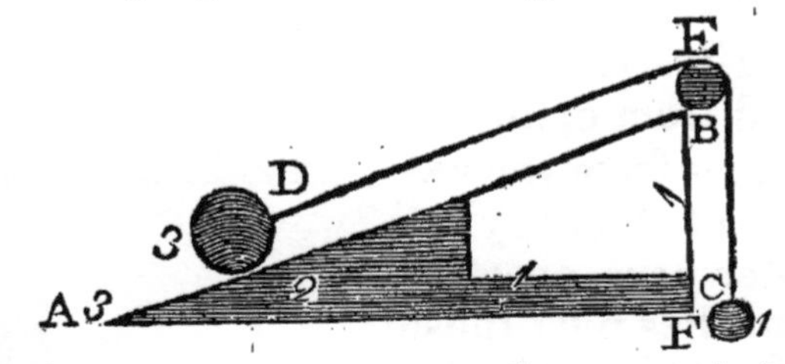

Thus the relative gravity, or the descending power of D, is in the same proportion to its actual gravity, as the height of the inclined plane C B is to the length of the horizontal plane A C.

Now, if B C be the perpendicular height of part o a hill, and equal to one half of D C and one third of A C, the horizontal planes from different sides of the road; a carriage may be drawn up with one third less power by crossing the road from A to B, than it would require in ascending directly from D to B. This is sufficient to show the great advantage loaded carriages obtain in ascending hills by crossing from one side of the road to the other; but here it must likewise be observed, as in every other part of mechanics, that what is gained by the power is lost in the time, inasmuch as the line A B is longer than D B.

The Wedge.

The wedge is a double inclined plane, whereby bodies are separated, (or rather the forces are in equilibrio) when the power applied is to the resistance as the thickness of the head or back of the wedge is to its length.

Illustration. Let A and B, two cylinders, be suspended from C, and let lines from the axis of each cylinder pass over the pullies D E; those from the cylinder B passing over D, and from A over E, each having a weight of two pounds fastened at F and G; then the cylinders will be drawn together, and form a resistance to the sides of the wedge A K and K B. Now, supposing the resistance of the cylinders to act with a force of two pounds each, and the length of the wedge B K to be equal to twice its thickness A B, the pressure of two pounds on the top of the wedge I will be equivalent to four pounds, the resistance from the sides of the cylinders.

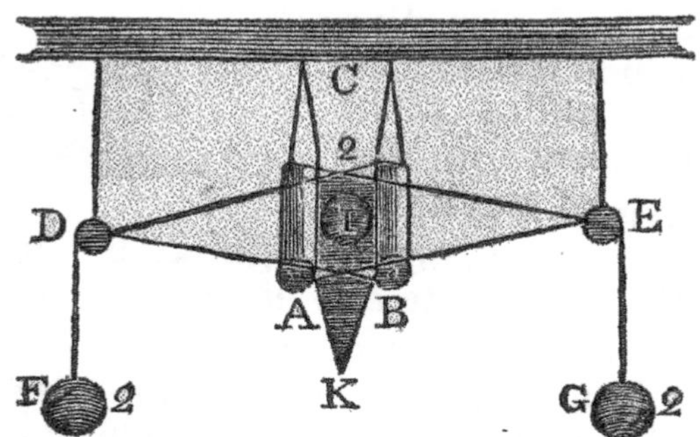

Remark. But the force that is given to a wedge is generally applied by a stroke, and not by the dead pressure of a weight; and a smart blow with a hammer of four ounces will overcome more resistance than a weight of two pounds laid on the top of the wedge. A blow from a hammer of fourteen pounds weight, will overcome more resistance in cleaving a log of wood, than the pressure of a ton weight laid on the top of the wedge. The suddenness of the blow, whereby the acquired momentum is expended all at once, causes a more easy separation of the cohesive parts, and the power of the hammer is greatly increased by the multiplication of its velocity into its weight.

Screw.

Remark. This machine is a spiral thread or groove cut ascendingly round a cylinder. When the spiral is formed upon the cylinder, it is called the entering screw; but when it is cut in the inner surface of a hollow cylinder, it is called the receiving screw. If the spiral thread were unfolded, it would form an inclined plane, the whole length of which would be to its whole height as the circumference of the cylinder is to the distance between the threads.

The screw is an inclined plane ascending spirally around a cylindric shaft, whereby pressure is produced, or heavy bodies raised; and the mechanical advantage is, as the length of one circular thread is to the distance between the threads.

Illustration. If the circumference D B be five inches, and it is required to place the threads at the distance of half an inch from each other, it is obvious that the spiral, in winding round the circumference, must pass through a line five inches long in attaining half an inch up the cylinder, which is the distance between the threads. Therefore, if the inclined plane G F H, which is half an inch in height and five inches long, be cut out and pasted on the cylinder, it will form one complete revolution of the spiral; or if E F and C F be five times G F and F H, the paper will roll five times round the cylinder, and the edge of the inclined plane will form a spiral or thread half an inch wide, and two inches and a half in extent.

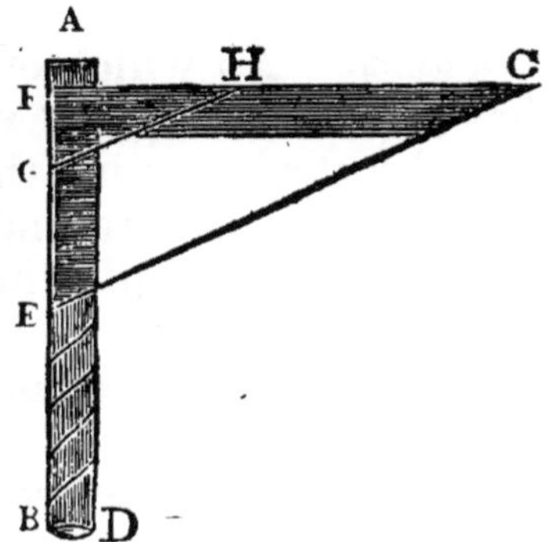

As the power of the inclined plane is as its length to its height, so the power of the screw is as the circumference of the cylinder to the distance of the threads.

The lever and screw are generally combined.

Thus, if the threads on the cylinder A B be made a quarter of an inch apart, and the nut D be turned by the handle or lever C D, which is two feet long; then as the power of the screw is relatively as the circumference is to the distance of the threads, this diameter of the power will be twice C D or 48 inches, and its circumference about 150 inches; so that the power is to the force or pressure as 150 to $\frac{1}{4}$, or as 600 to 1. Now, let 150*lb.* the ordinary force of a man, be applied to C, the end of the lever, and the pressure on the block B will be 600 multiplied by 150, or 90,000 *lbs.*

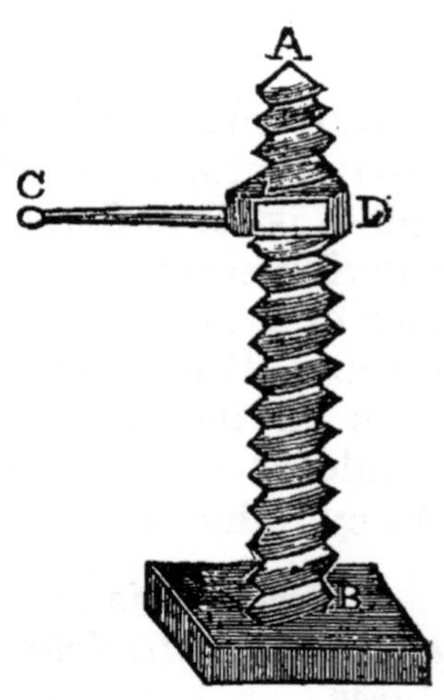

Query. To which of the mechanical powers does that power belong, by which heavy bodies are separated by straightening the joint of an interposed bar? This power has recently been applied to printing presses. We apply the same power in pushing bodies from us, when we effect their removal by merely straightening the previously flexed elbow.

Friction or Attrition.

Remarks. These calculations being made without regard to friction or attrition, which greatly affects all mechanical operations, a few remarks upon that subject may be profitable.

Friction is the resistance that a moving body meets with from the surface over which it passes. A carriage wheel as it turns on the road, is impeded by the inequalities on its surface; and the axle on which the wheel turns, likewise retards it by its attrition in the nave. The friction is much further increased when the wheels are locked or fastened, so that they drag upon the surface of the road; it is, therefore, to increase the resistance by augmenting the friction, that a wheel is locked in descending steep hills, where the relative gravitation gives too much velocity to the carriage.

Levers, axles, pullies, wedges, screws, and, in short, every mechanic power, or any description of body, is considerably retarded by friction, when it acts upon another body, either in motion or at rest.

Friction is considered as an uniformly retarding force in hard bodies, and not subject to alter by different degrees of velocity. It increases in a less ratio than the quantity of matter or weight of the body; and the smallest surface, or the fewest parts in contact, has the least friction, the weight being the same.

The force or power of friction varies in proportion to the different surfaces in contact; that is, accordingly as the surfaces are hard or soft, rough or smooth; even the hardest bodies which have the highest polish are not free from inequalities on their surface, which retard their motion when they act upon each other. When polished iron and bell-metal are opposed to each other in motion, they produce less resistance than bodies in general; but even these polished planes do not lose less than an eighth of their moving power, and others not less than one-third of their force, by friction.

As the friction between rolling bodies is much inferior to that which is produced by bodies that drag, the attrition of the axle in the nave has been lessened by a contrivance made with a number of small wheels, which are called friction rollers. These are held together in a box by its inflexed ends, and fastened in the nave, so that the axle of the carriage may rest upon them; and they turn round their own centres, following each other around in the box as the wheel continues its motion. A represents a section of the axle, C C the nave, and B B the friction rollers, which turn round their own moveable centres as the wheel revolves, and they progress round the axle of the carriage.

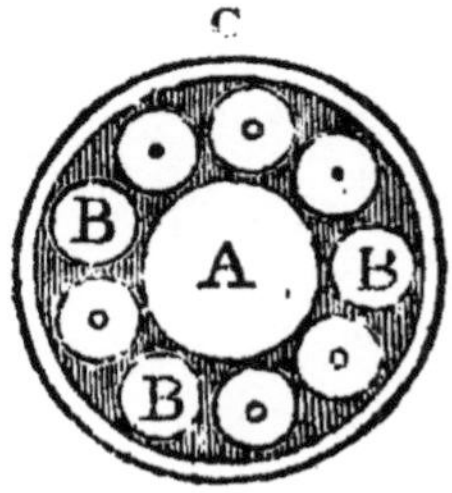

Cylindrical and spherical rollers are used with great advantage in turning heavy bodies, such as the top of a windmill, or the dome of an observatory; or in moving large logs of wood or blocks of stone from one place to another. The grand equestrian statue of Peter the Great, at Petersburgh, was formed out of an immense block of stone, which was brought from a place some miles distant, by rolling it along the road on iron balls laid on thick planks.

PNEUMATICS.

Before introducing the learner to this branch of Natural Philosophy, he should have a general knowledge of the principles of the Air-Pump, Barometer, Thermometer, and Hygrometer.

Air-Pump.

This ingenious and useful machine was the invention of Otto Guericke, about the year 1654; but it was soon afterwards greatly improved by Boyle, and has since been brought to a great degree of perfection by succeeding philosophers.

This machine is, of all others, the most useful in the prosecution of pneumatical studies. By means of the air-pump, we change the ordinary effect of the atmosphere; show the state of existence of bodies under different modifications of air, and how highly essential it is to the preservation of animal and vegetable life.

From the construction of the air-pump, it is impossible to exhaust the whole of the air; for the effect produced by the pump depends upon the elasticity of the air which is left in the receiver; this opens the lower valve of the machine, and the air escapes into the cylinders: therefore, when the air in the recipient has not sufficient elasticity to force open the valves, the action of the handle cannot produce any further rarefaction.

The construction of the air-pump is as follows:—

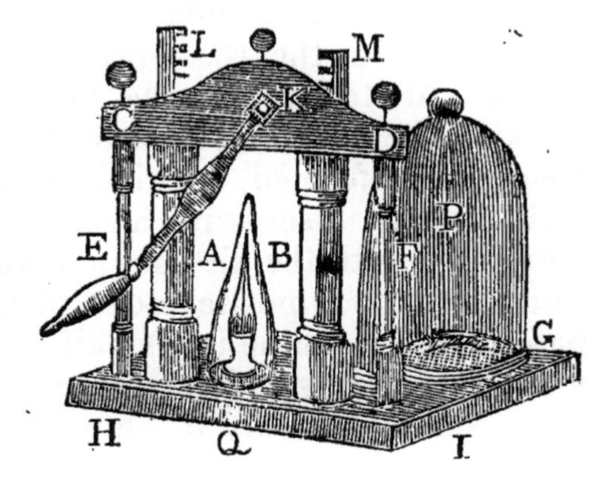

A and B are two brass cylinders, which are closely and firmly fastened down to the table or base of the machine H I by the head C D, and the columns E F; P is the receiver, which stands on G, a brass circular plate; this plate has a small hole in the middle, through which the air passes from the recipient into a closed channel made of brass, which communicates with the cylinders A B: near the bottom of each cylinder is a valve or lid opening upwards, and above these valves are two others, which are moved up and down by the toothed rods L M, that fall into a toothed wheel sunk in the block C D, to the axis of which K the handle is fixed.

On turning the handle, one of the pistons is raised and the other depressed, consequently a rarefied space is formed between the upper and lower valve in one cylinder; then the air which is contained in the receiver rushes through the conducting pipe, and by its elasticity forces up the lower valve, and enters the rarefied part of the cylinder L A; then the valve closes, which prevents the air from returning again into the receiver. When the motion is reversed, M the other piston ascends, and L is depressed; in its depression, the elasticity of the air contained between the two valves, forces open the uppermost valve, and it escapes into the upper part of the cylinder; then the valve closes again, and prevents its return.

The opposite piston performs the same operation, but the motions are alternate, so that whilst one piston exhausts the air from the receiver, the other is discharging it from the top of the cylinder. Thus, by continued exhaustion, the density of the air keeps decreasing in the receiver, till its elasticity is no longer able to force up the lower valves, which terminates the effect of the machine. When the experiment is performed, the air is again admitted into the recipient, by unscrewing a small nut at Q which communicates with the air channel, and restores an equilibrium to the opposite sides of the receiver. In the base of the machine is a small hole, which enters the air-pipe; and over this is placed a quicksilver gage, and a small receiver, to show the different densities of the air in the recipient when the machine is at work.

Barometer.

After the gravitating force or pressure of the atmosphere was discovered by Galileo, it was found by experiment, that water might be raised by a common pump to a certain height, and no further. He then happily imagined that this limited ascent must be the counterbalancing power to an equal column of the atmosphere. This idea was seized and improved upon by Torricelli, Pascal, and some others; who considered, that if the weight of a column of water 34 feet high, which is about the height it ascends in vacuo, be equal to a column of the atmosphere of the same base, any other fluid, differing in specific gravity from water, would balance the same column of the atmosphere, by rising to a proportionate height; and as quicksilver is about fourteen times heavier than water, it was supposed that a column of mercury of about the fourteenth part of the height of a column of water, would counterbalance the air like the column of water. The experiment equalled the wishes of the experimentalists; and after filling a glass tube with mercury, which was closed at the upper end to keep off the weight of the atmosphere, they found the mercury fluctuated, at different times, between the heights of 28 and 31 inches: this variation was discovered to arise from the different pressure of the atmosphere, which varied according to the different states of the weather. Thus a pneumatical instrument was formed, which has equalled, if not surpassed, any other in promoting physical knowledge. The barometer is variously constructed, for the purpose of accuracy and convenience. In the following descriptions, we have noticed those which are in the most common use.

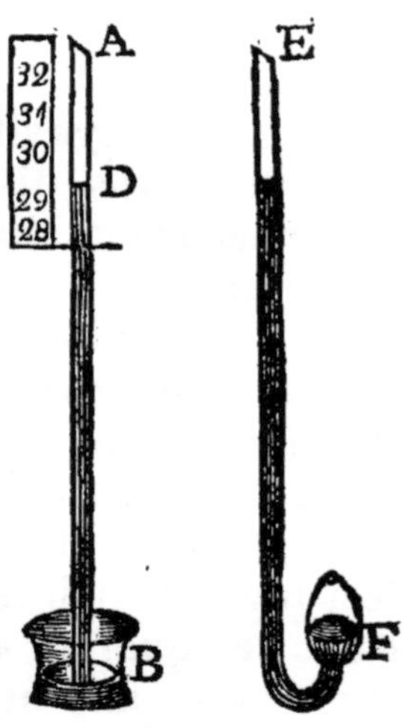

If A B be a small glass tube about 34 inches long, and a quarter of an inch in diameter, closed at one end; when this tube is filled with mercury that has been thoroughly freed from air,* and inverted in the basin, the mercury will sink down in the tube to the point D, somewhere between 28 and 31 inches from the surface of that which is in the basin, leaving a vacuum in the upper part of the tube: so that this part opposes no resistance to the rising and falling of the upper surface of the mercury, and leaves the free action of the atmosphere to press on the lower, without any counterbalancing force; which causes the mercury to rise in the tube as the density of the atmosphere increases, and to sink, as the pressure decreases on the surface in the cup.

Barometers are usually made of a tube, with a curved neck and bulb, being more commodious than the basin and tube. This is fastened to a frame, which has a scale of equal parts placed between 28 and 31 inches from the surface of the mercury, being the extreme variation of atmospherical pressure: and this scale likewise contains a prognostic state of the weather against different heights of the mercury, with a moving index and nonius to determine the changes, either in rising or falling.

To make these barometers tolerably exact, the circular area of the bulb should be at least 30 or 40 times larger than that of the tube, so that the mercury may be as little affected as possible whilst it rises and falls; if the mercury were to rise half an inch in the tube, and fall a quarter of an inch in the bulb, the difference of the height would be three quarters of an inch; for the height of the column is taken from the surface of the mercury in the bulb to its height in the

* By boiling it in the tube. *Patterson*.

tube. Therefore, according to the above variation, the surface of the mercury against the scale at the top of the barometer, would be a quarter of an inch short of its due height; but by increasing the circular area of the bulb to forty times that of the tube, the rising or depression of the mercury in the bulb at the greatest variation in the tube, would not be more than the tenth or twelfth part of an inch, which makes but an inconsiderable difference in the accuracy of the instrument.

Diagonal Barometer.

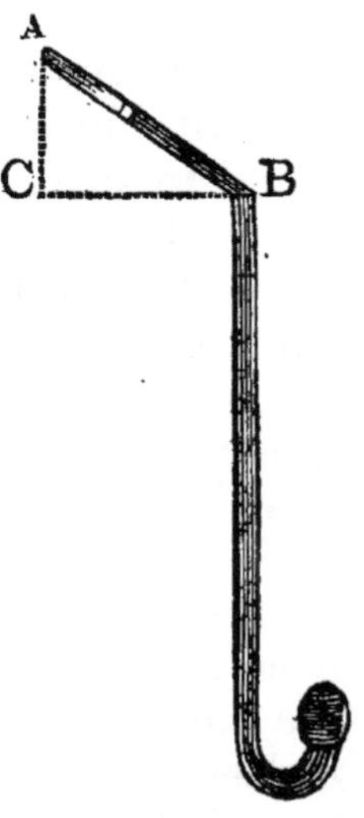

The inflected or diagonal barometer, which was invented by Moreland, is nothing more than the common barometer, with the upper part of the tube bent into an obtuse angle at B; the line A C, which is about three inches long, is equal to the perpendicular height of that part of the instrument, which is partly empty in those tubes that are straight, when the atmosphere is in a rarefied state; and as A B is longer than A C, the mercury must pass through a greater space in attaining the same height, consequently the variation, either in rising or falling, is more minutely determined.

There are many other kind of barometers, the construction of which renders them less perfect than the preceding; therefore, as their description would be more curious than useful, they have been omitted.

Thermometer.

The thermometer is a small tube and bulb filled with a fluid, and is used to determine the various degrees of heat in the atmosphere, water, spirit, &c. We find, by experiment, that air, and all kinds of fluids contract and expand by different degrees of heat; therefore, if any fluid body be enclosed in a small glass tube purged from air, it will sink or contract from exterior frigidity, and rise or expand on the increase of heat.

This tube of the instrument is generally filled with spirits of wine, linseed oil, or quicksilver. Spirits of wine, being very limpid, is very useful in observations on air, where there is no great difference of heat; but if the heat be too violent, the expansion of the spirit will burst the tube; and if the cold be extreme, the spirit will freeze; consequently, in either case, the instrument is useless. To remedy the first of these inconveniences, Newton filled his thermometer with linseed oil, which requires four times the heat of boiling water to cause ebulition: this was found to succeed very well in experiments of heat, but in those of frigidity it is much more susceptible of cold than the spirit thermometer. To meet both sides of the difficulty, Fahrenheit made one with quicksilver, which preserves its effect in either extreme.

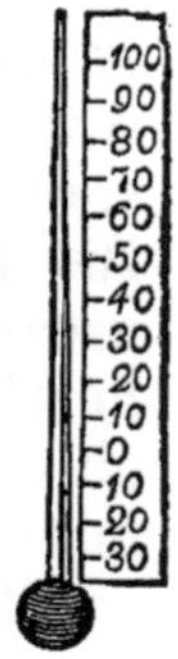

When this instrument is constructed, the bulb and tube are filled with a certain quantity of mercury that has been thoroughly purged from air or moisture by boiling; and before the end of the tube is hermetically sealed, or melted and closed, the air is entirely expelled from it by heating the mercury, which rarefies it, and drives it to the top of the tube.

A scale is annexed to the side of the tube, which Fahrenheit divided into 600 parts, beginning from the point (0), the greatest degree of cold which could be produced by surrounding the bulb of the thermometer with a mixture of snow or pounded ice, with sal ammoniac or sea salt. The point at which mercury begins to boil, he laid down as the greatest degree of heat, and the intermediate distance was divided into 600 parts, to show the various degrees of heat between the two extremes. When this thermometer was immersed in water just beginning to freeze, or in snow or ice which was beginning to thaw, the mercury stood at the 32d division of the scale; this he called the freezing point: and when the instrument was placed in boiling water, the mercury rose to 212, which he made the boiling point; making a difference of 180 degrees between these divisions.

But the more ordinary way of constructing thermometers at present, is to place them, after they are properly filled, in water just beginning to freeze, marking 32° for the freezing point; and then afterwards to immerse them in boiling water, and to mark the other point 212°, dividing the intermediate distance into 180 equal parts, for the scale of the instrument, which may be carried upwards or downwards to any greater extent, if it should be required.

Small portable thermometers, containing only a part of the scale, are constructed for those purposes where the variation is not considerable.

Hygrometer.

This is an instrument which is used to ascertain the different degrees of humidity or dryness in the air. There are many inventions for this purpose; but an instrument sufficiently accurate for common observation, may be made by fastening a weight to a piece of twisted catgut, and then suspending it from a nail in the wall; the catgut will contract in damp weather, and extend as the air becomes drier; and the difference may be shown by a scale of equal parts fixed on the wall near the weight.

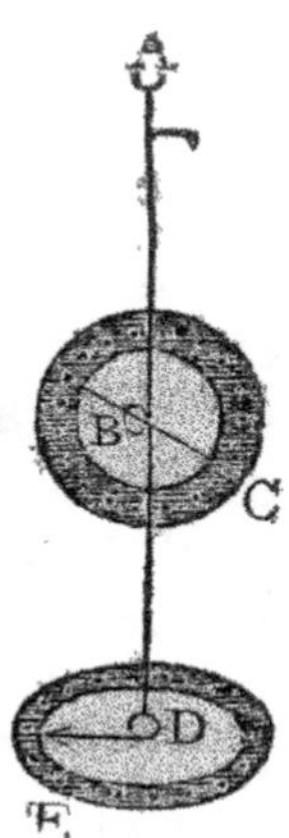

An ingenious improvement has been made to this method, by passing the catgut round a pulley and index at B, which points out the variation on a circle that is divided into equal parts, and fixed on the wainscot or wall. Another index may be fixed to the weight D, which, by the twisting and untwisting of the string according to the state of the atmosphere, will point out the changes upon a horizontal circle at E; thus a compound instrument may be formed, to correct its own operation.

A still better method, though not hitherto much in practice, was invented by Dr. John Torrey, and communicated to the editor when the inventor was a boy of 14 years. It consists of a piece of rattan, coiled as when wound on a whip-stalk. An index attached to one end of it, points to the degrees of moisture, &c. as the coil unwinds and winds up.

Remarks. Atmospherical or common air, which constitutes the chief subject of Pneumatics, is a rare, transparent, and elastic fluid, that surrounds the earth to a considerable height, and revolves with it in its diurnal and annual motion. Independently of light, heat, and electrical fluids, it may be considered as the common receptacle for the parts of all those bodies that are capable of being volatilized by heat, or dispersed by putrefaction, exhalation, evaporation, or any other principle that changes the animal, vegetable, or mineral productions to the aeriform state.

But the hand that formed this heterogenous compound, has likewise tempered it for the most essential purposes of life, throughout the animal and vegetable creation; so that it may be considered as one of the prime agents of nature, in perfecting and preserving her works.

Air is generally ranked amongst fluids; but it differs from liquids, in that it admits of indefinite density by increasing compression, and that it is incapable of fixity by any known degree of cold.

When the particles of air are acted upon by some other body, they move in every direction, and communicate sound, odour, or effluvia, to distances proportional to the given impulse.

Weight and Pressure of the Air.

Atmospheric pressure will sustain a column of water in a tube to any height, not exceeding about 34 feet.

Illustration. As it is difficult to introduce to the student a tube 34 feet long, mercury is generally used, as explained with the torricellian tube, because a 31 inch tube will do. But the principle may be illustrated by filling a narrow deep tumbler or assay glass with water, and then covering the open end with a piece of paper to even the surface; now invert it, while the hand is pressed on the paper: take off the hand carefully, and the water will be kept up in the vessel by atmospheric pressure.

The weight of the atmosphere compresses the animal body, and keeps the fibres from being forced out of their natural order; but as it presses equally on every part, we are insensible of its effects, except it be partially removed.

Illustration. Place the hand on the top of a small glass, called a hand-glass, which is open at both ends, and stands on the plate of the air-pump; then exhaust the air which it contains, and the fibres or fleshy part of the hand that cover the top will distend, and rise up in the glass, with a painful sensation, which is occasioned by the want of atmospherical compression on that part of the hand which covers the mouth of the glass. Whilst the hand remains in this situation, the weight of the atmosphere which presses on the upper surface will fix it so firmly to the top of the glass, that it requires a considerable exertion to remove it. If a piece of thin glass be placed on the top, instead of the hand, when the air is exhausted, the glass will be broken by the pressure of the atmosphere on its exterior surface.

As heat rarefies the air, the effect of the preceding experiment may be produced with a common wine-glass. Put a piece of burning paper into the glass, and after the air has been considerably rarefied by the flame, place the fleshy part of the hand evenly on the mouth of the glass, and the pressure of the atmosphere on the exterior part will press it so firmly to the hand that it will require some force to remove it. The pressure of the air is best shown by its effect on the barometer; it causing the mercury to rise and fall in the tube, as the weight of the atmosphere increases or decreases.

A portion of atmospheric air may be weighed in the scales, like other material substances.

Illustration. On weighing a glass bottle which contained 40 cubic inches, and afterwards exhausting the air and weighing it again, it was found to have lost 10 grains of its original weight, which is in the ratio of about 8 grains to a pint. The quantity of air exhausted out of the bottle was 34 inches; for on immersing the bottle in water in an inverted position, the quantity that flowed in and occupied the space of the exhausted air, weighed 8628 grains, which being divided by $253\frac{1}{3}$, the number of grains in a cubic inch of water, produces 34 inches for the quantity of air exhausted out of the bottle. Thus, it is found that the relative weight or specific gravity of air to water, is as 10 to 8628, or as 1 to $862\frac{4}{5}$.

Variation of gravity is caused in the atmosphere chiefly by the variation of temperature, and of the quantity of aqueous vapors.

Illustration. Set the barometer or torricellian tube in a cold room. After making a fire in it, the mercury will fall. When it becomes stationary, fill the room with vapor, by pouring water on a hot iron, and the mercury will fall still lower. In those parts which lie between the tropics, where the heat is constant and regular, the variations in the density of the atmosphere are likewise constant and regular; as the barometer is observed to sink about half an inch every day when the sun is above the horizon, and to rise again to the same point in the night. But from the tropics to the poles, the variation is irregular and inconstant, as the mercury is almost perpetually moving in the tube of the barometer from about the height of 28 to 31 inches; which serves to indicate the various changes that are likely to take place in the weather. Why these changes thus irregularly happen, still depends chiefly on conjecture. It is imagined that the currents of air which are formed by the inequalities of the earth as the whole atmosphere passes over its surface, give or receive various degrees of heat as the currents are rarefied or compressed, in passing by mountains, moving over plains, or descending from different heights of the atmosphere. The upper part of the atmosphere is found to be a region of perpetual frost. On the immense range of mountains, called the Andes, in America, as well as on many others, there is a very great difference of climate even at the same time. As the situation of one part of these mountains is almost under the line, it rests its base on burning sands; about half way up is a most pleasant and temperate climate, covering an extensive plain, on which is built the city of Quito, whilst the top is covered with eternal snow, perhaps coeval with the mountain itself.

In very dry cold weather, the atmosphere will sustain a column of mercury about 31 inches high—in fair settled frosty weather, 30½ inches—fair weather, with but little frost, 30—changable and cold, 29½—rain and snow, 29—hard rain and snow, 28½—stormy and warm, 28.

Atmospheric pressure holds down upon the table exhausted vessels which are open at the bottom, with great force.

Illustration. Upon the plate of the air-pump set a common receiver; the force or pressure of the air which is contained in the receiver being equivalent to that which acts on the exterior part, it may, like our bodies, be moved with facility. But as the air is exhausted, the equilibrium is destroyed between the interior and exterior surfaces, till the pressure of the atmosphere on the outside fixes the receiver so firmly to the plate, that it requires a greater force than that of one man to remove it. In treating of the barometer, it is there shown that the atmosphere presses with a weight of 15*lbs.* upon every square inch of the earth's surface; therefore, supposing the surface of the receiver to contain only 36 square inches, the pressure of the atmosphere upon it is 36 multiplied by 15, or 540*lbs.* weight; but some allowance must be made for imperfect exhaustion.

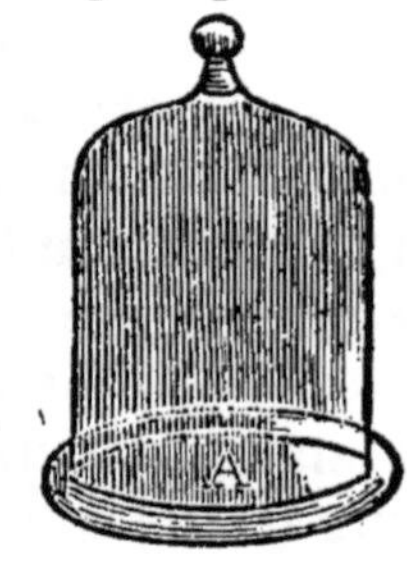

The rising and falling of the mercury in the pump-gage, shows the different degrees of density of the air in the receiver.

Illustration. B is a small tube filled with quicksilver, and immersed in the cup A, which likewise contains any given quantity of mercury; these are placed under C, a small receiver, which is set over the aperture of the air channel in the block of the machine, or it may be set within the common receiver. Now, before the exhaustion takes place, the mercury is supported in the tube B by the pressure on the surface at A; but as the air is withdrawn by the operation of the pump, the pressure on the surface of the mercury decreases as the density decreases; consequently that which stands in the tube loses part of the support that sustains it, and descends into the cup in proportion as the air is exhausted, which serves to show the density of the air that remains in the receiver.

Atmospheric pressure will burst some vessels inwards, and compress others together powerfully, provided they were previously exhausted of atmospheric air.

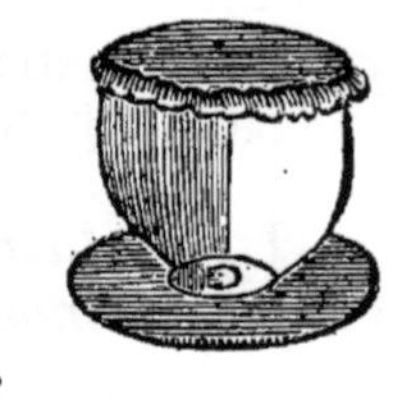

Illustration. Cover one end of a glass vessel which is open at both ends, with a piece of wet bladder, and leave it to dry. After it is perfectly dry, place the open end on the pump plate, and exhaust the interior air, till the weight of the air on the top of the bladder bursts it with a considerable report. If the air be exhausted out of a thin square glass bottle, the exterior pressure will break it to pieces.

A pleasing experiment, and a demonstrative proof of the gravity and pressure of the atmosphere, is shown by what is usually called the Magdeburg Hemispheres.

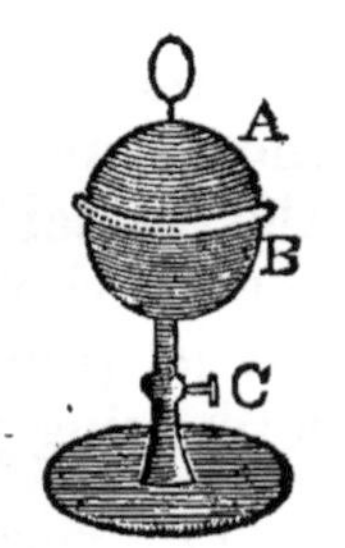

This machine is a hollow globe of brass, divided into two hemispheres A B; in the lower part C are a stop-cock and a tube which screws into the pump-plate. Before the air is exhausted from the interior of the globe, the interior and exterior pressures are equal, so that the parts may be separated with the greatest facility; but when the counterbalancing force is removed from the interior, and the stop-cock is shut to prevent the return of the air, the pressure of the atmosphere on the surface will compress the two parts of the sphere so closely together, that it will require more than an ordinary force to pull them asunder.

By means of this experiment, we are practically able to determine the actual pressure of the atmosphere on any given surface; for suppose that the mouth of the hemispheres contains 12 square inches, and that by a steelyard hooked to the ring at the top, it requires 180*lbs.* weight to separate the two parts; then if 180 be divided by 12, the result is 15*lbs.* which is the pressure on a square inch of the surface. This atmospherical pressure on bodies is confirmed by another experiment, to be described hereafter. If these hemispheres be placed under an exhausted receiver, so that the pressure of the air on both sides of the machine be made equal, they will separate with the greatest ease; which is an additional proof that the pressure of the atmosphere alone holds them together.

A fountain in vacuo may be formed, showing the force of atmospheric pressure on the surface of bodies.

Illustration. A tall receiver is placed over a brass plate and jet-pipe; these are connected with the air-pump by means of a stop-cock and tube: after the air is exhausted out of the receiver, the cock is shut to prevent its return; then the whole is unscrewed from the plate of the receiver, and the lower end of the tube is immersed in a vessel of water: on opening the stop-cock, the pressure of the atmosphere on the surface of the water in the vessel having no counterpoise from the interior of the cylinder, forces up the fluid through the jet-pipe with considerable velocity, which forms a pleasing jet d'eau, or fountain in vacuo.

The porosity of wood may be shown by atmospheric pressure.

Illustration. Let a solid piece of wood be fixed to the bottom of a cup, and passed, air-tight, into the neck of the bottle; when mercury is poured into the cup, and the air exhausted out of the bottle, the pressure of the atmosphere on the surface of the mercury in the cup, will force it through the pores of the solid piece of wood, and it will fall like a silver shower, to the bottom of the bottle.

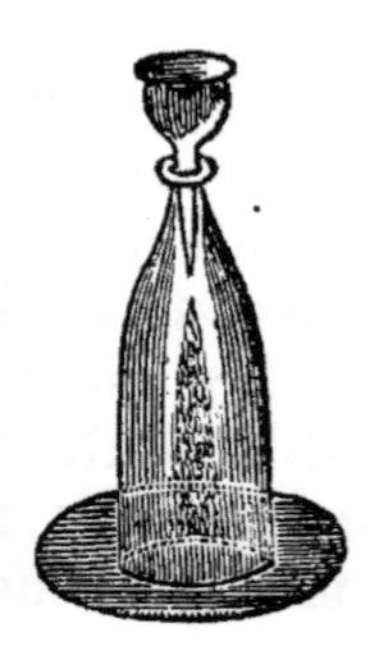

Sound is propagated by pulses of air, for it is scarcely audible under an exhausted receiver.

Illustration. If a bell be struck in vacuo, we are insensible of any vibratory effect; but as the air is admitted, the sound is augmented in proportion; so that if the density of the air in the receiver were increased beyond that of the atmosphere, the sound of a bell would be more forcibly heard in such a situation, than when it is struck in the open air.

The time of descent of light and heavy bodies in vacuo, is always the same.

Illustration. The difference of time which we observe in the open air, proceeds from the resistance of the medium through which they descend; but as this is nearly removed in an exhausted receiver, a feather will fall with the same velocity as a guinea.

Bodies which balance each other in the open air, may lose their equilibrium in vacuo.

Illustration. If a piece of cork and a piece of lead which balance each other in air, be weighed again under an exhausted receiver, the end that suspends the cork will descend; for when both these bodies are weighed in air, they lose the weight of an equal bulk of the air, consequently the cork loses more weight than the lead; but when they are placed under an exhausted receiver, what the cork lost by its magnitude in the open air, it now gains in vacuo; and as the bulk of the lead is much less than the bulk of the cork, the weight of the cork in vacuo will exceed the weight of the lead as much as their respective bulks of air exceed each other in weight.

The rise of vapour and smoke is caused by the density of the air.

Illustration. If smoke or vapour be placed under an unexhausted receiver, it will rise and darken the interior; but as the air is exhausted, the smoke descends, and at length leaves the vessel quite clear.

This serves to show that the air is lightest in moist and hazy weather; for then the density of the atmosphere is not sufficient to support the humidity it contains; therefore the weight of the vapour overpowers the resistance, and it descends in aqueous particles.

Winged animals are incapable of flight in vacuo.

Illustration. If a butterfly be suspended by its horns, or antennæ, from a thread in the middle of the receiver before it is exhausted, the insect will fly with apparent ease from one side to the other; but when the air is withdrawn, it hangs perpendicularly, and notwithstanding its efforts, it is unable to change its position.

Breathing, which is the principal action of life, arises from the compression and elasticity of air.

Illustration. If a small bladder be tied to a pipe and thrust into a bottle, so as to be air-tight in the neck, it may represent the lungs in the thorax, and the hollow tube the trachea or windpipe; then the air which is contained in the bottle about the bladder, shows the air in the breast that compresses the lungs, which is balanced by inspiration, or the air we inhale. When this is placed under a receiver, it shows the natural state of the lungs; but as the air in the receiver becomes exhausted, the elasticity of the air in the bottle begins to compress the bladder, till at length its sides are joined closely together.

The lungs of animals are compressed in the same manner when they are deprived of the counterbalancing force of the interior air; this produces a violent sensation or pressure on the breast, and stops the circulation of the blood in the lungs, which causes suffocation and death, unless the air be speedily admitted; then, if life be not too far exhausted, the sensation decreases, the action and reaction in the lungs become equal, and the animal recovers.

If a barometer be placed under a cylindric receiver on the plate of the air-pump, or a torricellian tube be extended upwards through the tubulature of a bell-glass, the mercury that stands in the tube will descend into the bulb, in proportion to the quantity of air which is exhausted out of the receiver, till the tube is completely evacuated.

Illustration. This shows that the counter-pressure of the air supports the mercury in the tube, and that the variation in the height of the column is caused by the different degrees of pressure on the surface of the mercury which is contained in the bulb.

The atmospherical variation is contained between 28 and 31 inches; therefore a mean distance of 29½ inches may be taken as the height and weight of a column of mercury, which is equal to a column of the atmosphere of the same base in a medium state. Now, as a cubic inch of mercury weighs about eight ounces avoirdupois, a square pillar of mercury 29½ inches in height, the base of which is an inch square, would weigh 15*lbs.* nearly, which are equal to the pressure of the atmosphere on a square inch of surface; these multiplied by 144, give 2160*lbs.* for the pressure on a square foot. Then, as the number of square feet on the surface of a middle-sized man is computed at 14½, this multiplied by 2160, produces 31320*lbs.* or nearly 14 tons weight, which perpetually presses on our bodies.

We may, by an easy calculation, find the pressure of the atmosphere on the whole surface of the earth, which is several millions of millions of tons.

Remarks. It may be asked, why we are not sensible of the great pressure of the atmosphere upon our bodies? To which it may be answered, that the pressure is counterbalanced by the air within us, which makes the sensation negative, and leaves us no more idea of the weight than we have of the motion of the earth, or of a ship on a smooth sea, when we are sailing at a great distance from the shore. But if this pressure be partially removed from any part of our bodies, we become sensible of it; for as the fibres are greatly distended and moved out of their natural order, it produces considerable pain in the part which is uncompressed.

Even the variation of the atmosphere generally produces a change in bodily sensation, which makes invalids very sensible of the alteration, without consulting their weather-glass.

The different densities of the atmosphere are of the greatest consequence to health; for in dull hazy weather, when the air is in its lightest state, the want of sufficient compression on the body produces pains in the breast, and a difficulty in breathing, which particularly affects those who have asthmatical complaints; it likewise suffers a distension of the various vessels in our bodies, which causes the circulation of the blood to lose part of its activity, and thus the whole system is oppressed with pain, languor, and debility. Mistaking causes for effects, it is a vulgar opinion that in dull hazy weather the air is in its heaviest state, but the reverse is the fact; for then the air is the lightest, which suffers the humid particles that float in it to descend to the earth.

In clear and serene weather the atmospheric pressure is the greatest, which braces or constringes our bodies, and gives such force to the circulation of the blood as removes obstructions in the vessels, and produces a proper tone in the system; thus tending to promote the blessing of health and the enjoyment of life.

The Elasticity of the Atmosphere.

Remarks. If the atmosphere which surrounds the earth were nonelastic, or of uniform density, like water, or fluids in general, its height might be easily ascertained by means of a barometer. For if the density of air be to that of quicksilver, as 1 to 11040, and if a column of quicksilver 2½ feet high, be equal in weight to a column of the atmosphere of an equal base, the whole height of the atmosphere, supposing it to be of an equal density, would be 11040, multiplied by $2\frac{1}{2}$, or 27600 feet, which is little more than $5\frac{1}{4}$ miles. But the air is found to possess an elastic quality, which gives it a density, in proportion to its compression, and causes the atmosphere to extend to an unlimited height. Some idea of the elasticity of air may be conceived by compressing a sponge or piece of wool, the fibrous parts of which distend in every direction as the pressure decreases.

The spaces which air occupies when it is compressed by different weights, are reciprocally proportional to the weights themselves; for the more air is compressed, the less space it takes up.

Illustration. Pour a small quantity of mercury into a bent pipe or tube A E, and it will rise to D; then stop it up at A, so that no air may escape from that part of the tube. When it is in this situation, it is evident that the weight of a column of the whole atmosphere, equal in diameter to the width of the tube, rests upon D the surface of the mercury. Now fill D E with quicksilver, till it is equal in weight to a column of the atmosphere, or about 29 inches high; then double the weight of the atmosphere rests on the mercury at D, which will force the fluid to B, half way up the opposite tube. From this it appears, that the space occupied by a certain quantity of air, under different pressures, is reciprocally proportional to the force of the pressures. Therefore, as the pressure of the upper parts of the atmosphere upon the lower becomes less according to the different heights, it must follow, that the air in the higher part of the atmosphere, where the pressure is very inconsiderable, may be rarefied to an almost unlimited extent.

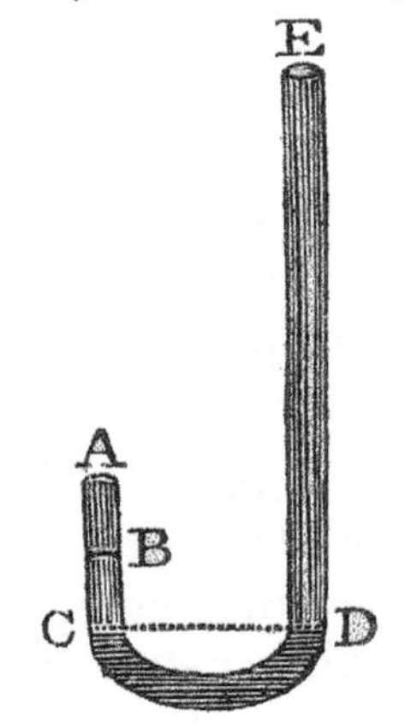

Now, supposing the atmosphere to diminish in density exactly in proportion to the different heights, the relative density of the air may be found at any given height. Thus, it is calculated that at the height of 3½ miles, the atmosphere is about twice as rare as on the surface of the earth; and at 7 miles, four times as rare, and so on in proportion, according to the following table:

Height in miles.	No. of times as rare.
3½	2
7	4
14	16
21	64
28	256
35	1024
42	4096
49	16384
56	65536
63	262144
70	1048576

Thus, pursuing this calculation, it would seem that a cubic inch of the common air which we breathe, would be so much rarefied at the height of 500 miles as to fill a sphere equal in diameter to the orbit of Saturn.

Remark. The elastic property of air differs from the elasticity of bodies in general. When solid bodies are compressed, they have an elastic power, which causes them to resume the same figure they possessed previous to compression; but on removing the pressure on air, it will not only resume its first bulk, but expand to any extent, as we have already described.

The elastic force of air acts in right lines; so that when the compressive power is removed, it diverges in all directions as from a common centre.

Illustration. This is evident from the small globules which are formed on the surface of an egg, or any other body immersed in water, when the exterior air is exhausted. When these globules first appear, they are exceedingly small; but as they increase in bulk, they still preserve their spherical form. Soap bubbles, or glass globes, are formed by the elasticity and equal divergency of the particles of air.

The elastic principle of air may be so applied as to project water enclosed in a vessel with great force.

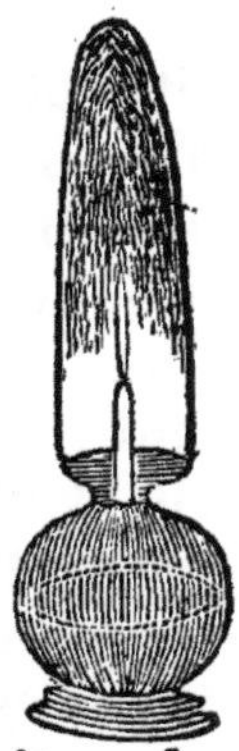

Illustration. If a glass bottle, with a small tube in its neck, extending to near the bottom, be half filled with water, and placed under a receiver; when the air is exhausted out of the recipient, that which is contained in the bottle loses its counter pressure, distends by the elasticity of its parts, presses on the surface of the water, and forces the fluid through the neck of the bottle, to a a considerable height.

The power of a fountain of this kind may be greatly augmented, even without placing it under an exhausted receiver, by forming the vessel of brass instead of glass, and by using an injecting machine to compress and condense the air in the upper part of the globe; for if the air be compressed in a brass globe, the sides of which are capable of resisting the expansive force of the injected air, the height to which the water ascends, will be proportional to the density or elasticity of the air that presses on the surface of the fluid: now, as the density may be made many times greater than that of the atmosphere, the elastic power which arises from the compression, will overcome the resistance of the atmosphere, and the water will spout out of the neck with a force proportional to the difference of the densities between the internal and external air.

Those who have not the advantage of a complete apparatus, may, by means of a phial and small tube, or a tobacco-pipe, produce a sufficient effect to satisfy themselves of the elasticity of air.

Fill a phial about half full of water, insert one end of the pipe in the fluid, and let the other project about an inch above the neck of the bottle; then close up the pipe in the neck with sealing-wax, so that the air may not escape from the bottle. After the machine is completed, blow strongly through the tube, and the elasticity of the air, which is compressed in the upper part of the bottle, will so far overcome the resistance of the atmosphere or exterior air, as to force the water out of the pipe some inches in height, till the density of the interior and exterior air become equal. When the water is exhausted below the end of the pipe in the bottle, it may be supplied by sucking the tube with the lips, and instantly stopping the aperture of the pipe with the finger; then immerse the end in a basin of water, and when the finger is removed, it will flow into the bottle. For as part of the air has been drawn out of the phial by the lips, that which remains is less dense than the exterior air; so that the pressure on the surface of the water in the basin overcomes the resistance of the rarefied air in the bottle, and forces the fluid up the pipe, till the gravities of the interior and exterior air become equal. As heat distends the volume of air, by imposing a superior degree of elasticity; if the phial be held near the fire, or even warmed by the heat of the hand, this will increase the elastic force of the air, and cause a small discharge of water from the neck of the tube.

The elastic principle of air will distend collapsed vessels, or even burst them, when external resistance is removed.

Illustration. If a very small quantity of air be tied up in a bladder, when it is placed under a receiver, the sides of the bladder will gradually distend, as the exterior air is decreased, till it is completely inflated by the elasticity of the air which it contains. If a superior quantity of air be left in the bladder, when the receiver is exhausted, the expansive force or elastic power of the air which is tied up, will burst with a considerable report.

Bodies in general contain a quantity of air, particularly wood, fruits, and other vegetables. If a shrivelled apple be placed under an exhausted receiver, it will be plumped out, and appear quite fresh, by the spring of the air which is contained under its skin. If the apple be immersed in water, under an exhausted receiver, part of the air which it contains will issue in small globules from every part of its surface. The air which is contained in wood or the pores of bodies in general, may be seen by immersing them in a similar manner. From the porosity of an egg-shell, the included air that is forced from the egg through the shell by the elasticity of the air, will form itself into beautiful pearly globules all over the surface: these globules of air are driven in again when the air is admitted into the receiver.

Weights may be raised by the elasticity of air.

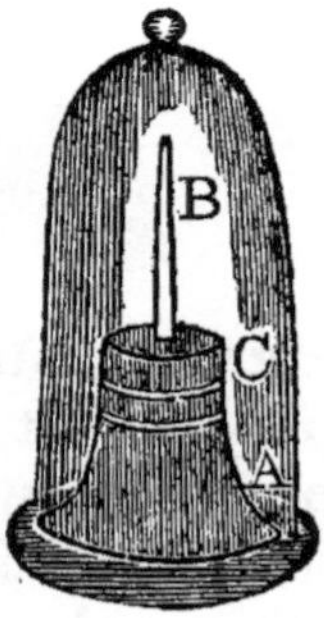

Illustration. The bladder A contains a small quantity of air, and is placed in the frame B, with the weight C laid upon it. On exhausting the air out of the receiver, the small quantity which is contained in the bladder will distend with such force, by its elasticity, as to raise up the weights that are laid upon it. By injecting air into cased bladders, with a forcing piston, any quantity of power may be obtained for raising considerable weights.

The expansion of air may be made to raise bodies in liquids of great specific gravity.

Illustration. Take a small bladder which contains a portion of air, and sink it with a leaden weight in a vessel of water; then place it under an exhausted receiver, and the elasticity or dilatation of the air, which it contained in the bladder, will raise it and the weight to the surface of the water. Balloons ascend by a similar principle; for the gravity of the air which they contain being less than that of the atmosphere, they rise in proportion to the difference.

Minute globules of air contained in water, by their elasticity, become enlarged, and ascend when atmospheric pressure is taken off.

Illustration. Set a florence flask of water, filled to the narrowest part of the neck, under the common receiver. On exhausting the air, a great number of large air-bubbles will ascend for a long time.

Air may be so compressed, as to project bodies with great force when liberated.

Illustration. An air-gun made of brass, and having two barrels, one within the other, the interior barrel receiving the ball like a common gun, and the exterior being filled with compressed air by a forcing syringe in the stock, may be made to project a ball to

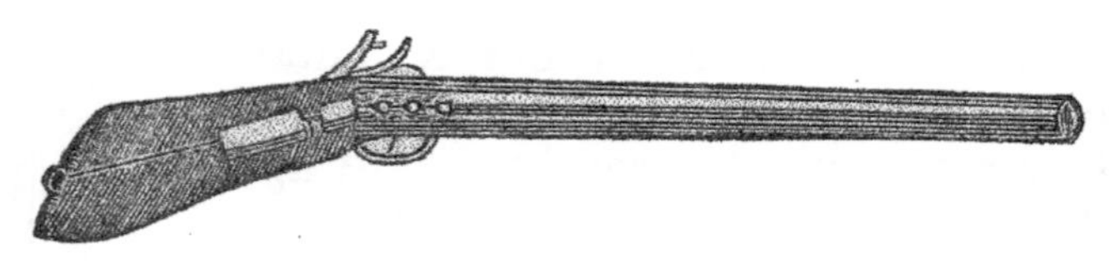

a great distance. After the air is sufficiently compressed, it is retained by shutting a valve that is placed between the syringe and air cylinder. When the trigger is drawn, the fall of the cock opens another valve, which communicates with the interior or ball cylinder; then the compressed air rushes in with such force on the back of the ball, that it drives it with great velocity out of the barrel. As the ball-valve shuts instantly, a number of shot may be discharged by one operation of the syringe; but as the compressed air loses a part of its force by every discharge, the velocity of the balls will regularly decrease, till the compression is renewed.

Remarks. The use of the diving bell is subject to the laws of atmospheric elasticity, though it can hardly be entitled to a place under this head.

The diving bell is used for the purpose of descending to great depths in the sea; and by means of this machine and its apparatus, the persons who are contained in it receive a supply of air for their existence, whilst they fasten ropes to cannon or packages which are on board sunken vessels, or to heavy bodies of any description at the bottom of the sea.

There are many forms of this machine, which have different advantages according to the different purposes for which they are employed. The following was invented by Halley, who made the first improvements in this hazardous machine.

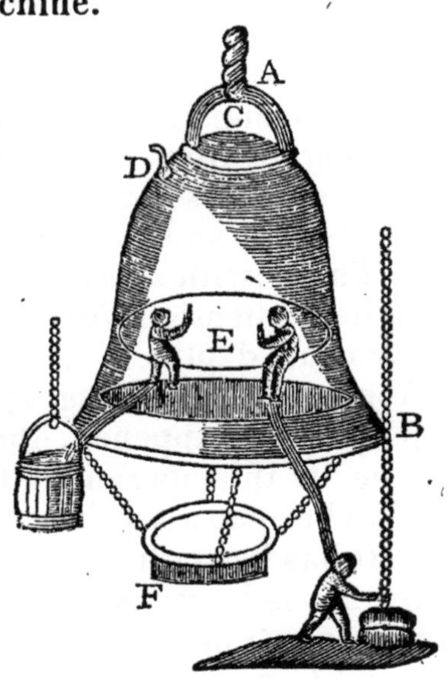

It is formed like a bell, and is about three feet wide at the top, five feet at the bottom, and eight feet high; containing about 63 cubic feet, or eight hogsheads. Its sides are loaded with lead to make it sink in the water, and on the top of the bell C, is a thick clear glass, to give light to the machine when it is immersed: D is a stop-cock, by which the impure and rarefied air is discharged: towards the middle E, is a seat for the divers to rest upon, and a broad iron rim F, is suspended by lines from the bottom of the bell, for the men to stand upon as occasion may require. When the divers leave the bell, they have strong globular caps, with flexible tubes, fastened to their heads; these caps have circular glasses in front, to give light; and the end of the flexible tube is kept in the bell, to supply the divers with air, whilst they fix tackles to those bodies that are to be raised. The bell is supplied with air during the whole operation, by barrels loaded with lead, which contain about 63 gallons each. The lower part of the barrel is open, and the water compresses the air in the upper during its descent. A close and flexible pipe, with a cock at the end, is fixed to the top of the barrel, so that when it has reached the side of the machine, the tube may be drawn in, and the cock turned, to admit the air into the bell; which is instantly effected by the great pressure of the water upon the air in the open part of the barrel; so that by a constant supply of air barrels, the divers are enabled to remain some hours under water. The corrupt or rarefied air in the bell being compressed by that which is fresh and more dense, rises to the top of the machine, and by opening the venting cock, it is forced out into the sea. There are likewise signal ropes connected with the machine, for moving it from one situation to another; and any particular instructions are scratched on a piece of sheet-lead, and sent up by a cord.

Thus, with a proper machine and apparatus, and vigilant attention from those who wait on the surface, even the depths of the sea cannot hide its treasure from the enterprising spirit of man; but it has too frequently happened, that by carelessness or misfortune in this operation, the lives of the adventurers have become a sacrifice to their undertaking.

It has been already shown, that the space occupied by air is in reciprocal proportion to the compressing power. Therefore, if a column of the atmosphere be equal to a column of water of the same base, 34 feet in height, the weight of twice that column of water would compress or overcome half the atmospherical column; consequently, if the body of air which is contained in the bell on the surface of the water, be let down to the depth of 31 feet, it will have a double weight acting upon it, which condenses the air, and suffers the water to rise half way up the bell; at 68 feet, the water will rise up two-thirds, and so on in a reciprocal proportion. But by means of the supplying barrels, the divers not only receive fresh air for their existence, but the compressive power of the water, which acts in the open part of the barrel, drives the condensed air with such force into the bell, as overbalances the compression of the water, and drives it out of the machine. Dr. Halley says, that he has effected this to such a degree, that he has been at the bottom of the sea in the bell, when the water did not rise over his shoes.

Vibrations producing sound.

Remarks. Sound arises from a tremulous or vibrating motion in elastic bodies, which is caused by a stroke or collision, and carried to the ear through the medium of the air. Thus the production of sound depends on three circumstances, viz. a sonorous body to give the impression, a medium to convey it, and the ear to receive it.

Sonorous bodies, such as gold, silver, copper, iron, brass and glass, produce strong sounds, in proportion to their density and elasticity. Lead, wood, wax, and other soft bodies, which want density or elasticity to give them any considerable vibratory motion, produce heavy, dull, and imperfect sounds.

Among the most agreeable variety of sounds, we may reckon the vibration of bells; the tones of which vary in proportion to their magnitude. Glass, from its elastic and vibratory quality, produces not only powerful, but exquisitely harmonious sounds. The strings of instruments, such as the harp, violin or harpsichord, produce a pleasing variety of sounds: these tones are varied by the length, thickness and tension of the strings or wires that compose the instrument. The condensation of air in the mouth of a flute, or the pipes of organs, &c. produces pulses of air or sound, independently of the elastic vibration of the body.

There are different opinions with respect to the manner in which the vibratory impulse is conveyed to the ear; but that it is effected by the medium of the air, is experimentally proved by the different degrees of sound that are produced in different densities of air under the receiver of an air-pump, which we have already explained.

Though air is the most usual vehicle by which sound is conveyed to the ear, yet it is not less powerfully transmitted by water, or a continuity of hard and sonorous bodies, where the air can have no possible operation. Dr. Franklin imagines, that with his ear under water, he has heard the collision of stones in that medium at the distance of a mile. The scratching of a pin on the end of a piece of hard timber, may be distinctly heard at the distance of 15 or 20 feet, by placing the ear at the opposite end.

To give a more forcible idea of vibratory motion, and its effects on the air in conveying sound: If a string be pulled tight, and be afterwards drawn on one side by the hand, on removing the power, it will pass over to the opposite side to nearly the same distance from its first position; thus it will continue its motion backwards and forwards, gradually decreasing the distances, till the vibratory motion ceases, and leaves the string in a straight line. According to the first law of motion, this vibration would continue perpetually, if it were not impeded in its progress by the density of the air, and the cohesive power of the string; but these resistances gradually shorten the distance of each vibration, till at length the impetus is destroyed. Now, as particles of air, like fluids in general, act in all directions, on the first impulse of the string, motion is communicated to all the surrounding particles, as if the sound were generated from a point; which is the cause why sounds are equally heard on all sides of the object of percussion, when the air is not acted upon by any other impulse.

The mode by which sound is communicated, may be shown thus: When the finger is removed from the string, all the particles of air that lay before it are driven forwards and condensed, whatever may be the velocity of the vibratory motion: this condensation acts as an impetus on those particles that are still further off, and gives a continued motion to the vibratory impulse, till it reaches the ear. When the density of the impressed particles is relaxed by the return of the string, it gives the same kind of vibrating motion to the air as that which it received from the string; for as the particles at a distance receive their compression and relaxation from those that are before them, the waves or pulses of the air cannot go backwards and forwards together, which would prevent the alternate condensation and rarefaction; but the pulses are agitated in different times, therefore they meet one another, and form the compression by which the sound is transmitted.

Sounds produced by the vibration of strings are on a high or low key, and quick or slow, according to the length of the strings.

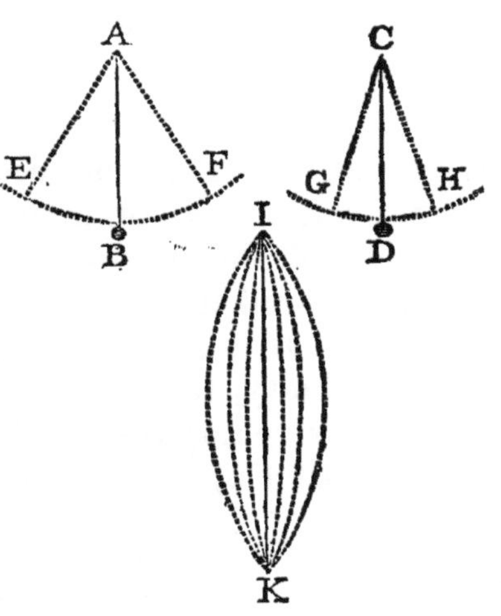

Illustration. According to the laws of pendulums, we find that those which are of equal length move in equal times, although they pass through different arcs; that is, if the pendulum A B and C D be equal, the time of passing through E F is equal to that of passing through G H. Thus, the vibration of the string I K is considered as a double pendulum, oscillating from the points K and I, the respective vibrations of which, from the greatest to the least, are performed in the same time. This is the reason why a musical string has the same tone from the beginning of the vibration to the end.

The vibratory motion of bodies that causes the pulses or undulatory motion in the air, which produces sound, is distinctly seen in extended strings, when they are drawn out of a right line, and left free.

Bells, and other sonorous bodies, vibrate like strings.

Illustration. When the side of a bell is struck by the clapper, it becomes elliptical from the elasticity of its metal, and has its greatest diameter from the point of concussion to the opposite side; this decreases as its vibratory motion decreases, till it again resumes its circular form. Although the motion of the metal may not be clearly perceived, the effect may be observed by throwing a small piece of paper on the surface of the bell, which will be considerably agitated during the vibration.

The power of vibratory motion, and the transmission of sound, may be sensibly felt, and communicated to the ear by touch.

Illustration. Take about a yard of riband or string, and tie it in the middle to the top of a small bar of iron, or a common poker; then twist the extremities of the riband round the forefinger of each hand, suspend the bar, and place a finger in each ear: when the bar is struck in this situation by some other sonorous body, the vibration will be forcibly felt, and the successive impulses will be heard with a force and tone like that of a large bell.

The pulsations of sound are radiated through the air in every direction, as from a given point.

Illustration. To make this more clear, the pulses have been assimilated to the small waves, or the undulating motion of water, which is formed in concentric circles, by throwing a stone into a standing pool; but as this representation has its generating circles formed in a horizontal plane only, perhaps it may be more aptly represented by any spherical-coated substance like an onion, the shells of which may represent concentric spheres or pulses issuing and diverging from a common centre.

Sound possesses equal velocity, whether it be soft, or loud, sharp or dull.

Illustration. Thus the tone of the smallest string will reach the ear as soon as that of the thickest, the softest whisper as soon as the loudest voice, or the report of the smallest pistol in the time of that of the largest cannon; but the distance to which the pulses are carried, depends on the impetus or force of concussion.

Decrease of sound is occasioned by the want of perfect elasticity in the air.

Illustration. If the action and reaction of the air were perfect, sound would continue to an infinite extent; but as every following particle has not the whole motion of the preceding, the further the sound passes the greater is the impediment it receives, from the want of free elasticity in the air; consequently the condensation of the pulses decreases till the impression is entirely lost.

According to the opinions of Derham and others, sound passes through the distance of 1142 feet in a second of time; and its audibility decreases in proportion to the squares of its distance from the object of vibration, or according to the extent and rarefaction of the concentric shells or pulses of air that surround the point of collision.

The distance of a sonorous body may be ascertained very nearly, by calculating the time which elapses between the pulsation and the reception of the sound.

Illustration. Although the velocity of sound is a little impeded or accelerated by currents of air, the distance from the object that generates it, may be nearly determined. Suppose the report of a cannon or of lightning, be heard five seconds after seeing the flash, multiply 1142 feet, the velocity of sound in a second, by 5, this will give 5710 feet, or rather more than a mile for the distance of the observer from the cannon, or place of electric explosion. As the velocity of light is considered as instantaneous, the flash is taken for the first moment of sound.

Remarks. Pleasing tones may be produced, which are denominated *musical sounds.* It is necessary that these tones should follow each other in a certain order of succession, to be pleasing. The tone of a sound depends on the time that the impression dwells on the ear, or the time that the string vibrates. Thus the longest strings have the longest vibrations, and produce the gravest sound; on the contrary, the shortest strings have the shortest vibrations, therefore occupy less time, and have the sharpest sound. The tone of the same string is equal through the whole of its vibrations, from the greatest to the least, as we have already stated.

The time of vibration which produces different tones, depends on the length, magnitude, and tension of the strings.

Illustration. If A B be equal in magnitude and tension to C D, and their lengths be in the proportion of 2 to 1, the times of vibration will be in the same proportion; that is, whilst A B makes one vibration, C D passes through two, or the vibrations coincide at every second of the shorter string. If the strings E F and G H be of equal lengths, and have the same tension or weight at N; but if one be double the thickness of the other, the time and number of vibrations in the smaller string will be double those in the thicker.

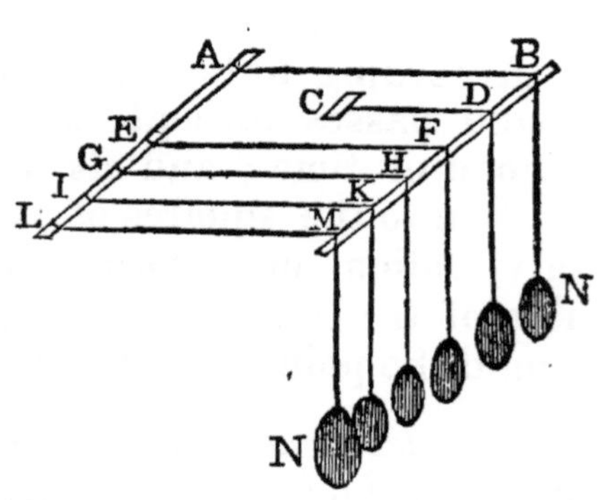

Or, when the strings are of equal length and thickness, but of different tension, the difference of time or vibration will be inversely as the square root of the weights N N, &c.; that is, if the weights are as 1 to 4, the times of vibration will be as 1 to 2, the square roots of 1 and 4.

In wind instruments, the vibrations or pulses will be in proportion to the length and width of the tube that compresses it.

Illustration. As the sound in wind instruments is produced by the elasticity of compressed air, the vibrations are longest where there is most room ; consequently the sounds are deepest in large and long tubes.

As the vibrations of strings coincide at different intervals, the more frequently this coincidence happens, the more agreeable is the sensation which is produced in the ear.

Illustration. The vibrations uniformly coincide when the strings are of the same length, magnitude and tension, which produces perfect unison or concord. The next greatest number of coincident vibrations, is when the strings are in the proportion of 2 to 1 ; that is, when the shorter string is half the length of the longer, and makes two vibrations whilst the shorter makes but one : these are called octaves or eighths.

If the vibrations be to one another as 2 to 3, the coincidence will be at the third of the shorter string, and in music it is called a fifth. If the vibrations be as 3 to 4, they produce a fourth, and so on through all compound vibrations, which form concords and discords, accordingly as the vibration of the different strings relate to one another.

Remark. When two strings of equal tone are placed near to one another, on striking one, the pulse or undulatory motion of the air will produce a sympathetic sound in the other. In like manner, strings of different lengths which are in concord with each other, communicate a vibration throughout the whole when any individual string is put in motion.

Sounds are augmented by being conveyed through a metallic or other tube, having a hard and smooth inner surface.

Illustration. The speaking trumpet augments sound, by the reflection of the pulses on the sides of the tube as they are propagated by the mouth. The aerial pulses, which are thus driven through the tube, not only augment the sound by increasing the aerial density of the pulses, but also by directing them more immediately to the object; likewise the reflection of the pulses on the sides of the trumpet receive additional force from the elasticity or reverberation of the metal; or rather, every point of percussion may be considered as a part from which fresh pulses are perpetually generating.

If a tube be continued to prevent the dispersion of the pulses, sounds may be carried to a very considerable extent; even the softest whisper may be distinctly heard at the distance of 15 or 20 feet. The whispering tubes sometimes surprise and amuse, when the communication is concealed, and the ends of the tubes terminate in the mouth and ears of two figures set at some distance apart.

Sounds are reflected by solid surfaces; by which reflection the sounds are repeated, or produce echoes.

Illustration. In the transmission of sound, it is conceived that every point of impulse serves as a centre for generating fresh impulses in every direction, and that sound passes through equal distances in equal times. Then, if the sum of the right lines with their reflections be equal to one another, the times will be equal; that is, if the pulses which diverge in right lines from a given point, be variously reflected on different sides, the sound will return in equal times to the generating point, or to any other where the distances become equal. This return of the pulses is called an echo.

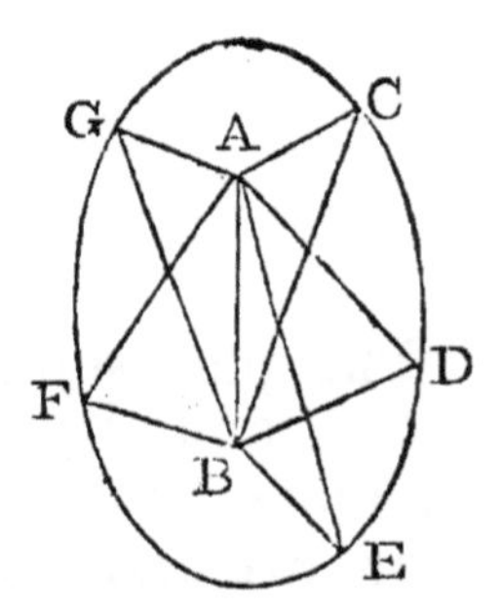

If aerial pulses be propagated from the point A, and strike various points of the curve CDEFG, and the sums of the respective lines taken together at B, be equal to one another; that is, AC+CB=AD+DB, and AD+DB=AE+BE, &c. then the echo or reverberation of sound will be heard at B, as a common point formed by the equal distances or times of the respective quantities of sound.

Sounds that follow one another are not distinctly heard, if they exceed 9 or 10 in a second of time; and as sound passes through 1142 feet in a second, the pulses of sound must precede each other by $\frac{1}{9}$ of 1142, which is about 127 feet, to be heard distinctly in succession. So that if the various distances through which the sound is propagated do not exceed A B by 127 feet, the echo will not be formed clearly at B. If the sums of the lines do not exceed each other by more than 127 feet, the sound which is reflected from the different points of reverberation will arrive so near the true time, that the difference will not be perceptible to the ear.

LIQUIDS.

The philosophy of liquids is divided into Hydrostatics and Hydrodynimics. The instruments or machines by which the principles of the latter are applied, are called Hydraulics.

Hydrostatics.

Remarks. This useful and interesting part of Natural Philosophy treats of the pressure and equilibrium of liquids.

By the word liquid, is meant a body, not of a gaseous form, compounded of small particles, which easily give way to an impressing force, varying their place and mixing with one another with great freedom and celerity. The constituent parts, or the particles which form a liquid, are conceived to be exceedingly small, smooth, hard and spherical; possessing the same nature and qualities as belong to bodies in general. A liquid is considered more or less perfect, as the particles which compose it move amongst themselves with more or less freedom and celerity.

Liquids are not perfectly dense, as a quantity of salt may be dissolved in water without augmenting the bulk of the water, and another salt may be dissolved in its saturated solution. This leads us to imagine that the particles of water are spherical, and that the interstices which are formed between them are occupied by the salt, in the same manner as when fine sand is poured into a case of shot, which fills up the vacuities without augmenting the bulk.

Liquids, like most solid bodies, change their appearance by the different modifications of heat. Water is somewhat viscid, as it will form bubbles; but a superior quantity of heat destroys the cohesive force of the particles, and forms them into vapour; and an inferior quantity increases the cohesion, and forms them into a solid mass of ice.

This changeable quality, which belongs to bodies in general, leads us to suppose that all the particles of matter are constituted alike, and that the different appearance of bodies arises from the various modifications of the particles which compose them. Be this as it may, one common property is clear: that all bodies, whether solid or fluid, consist of heavy particles, the gravity of which is always proportional to the quantity of matter which they contain.

All liquids are incompressible in any considerable degree.

Illustration. The academy del Cimento, from the following experiment, supposed water to be totally incompressible. A globe made of gold, which is less porous than any other metal, was completely filled with water, and then closed up. It was afterwards placed under a great compressive force, which pressed the fluid through the pores of the metal, and formed a dew all over its surface, before any indent could be made in the vessel. Now, as the surface of a sphere will contain a greater quantity than the same surface under any other form whatever, the academy supposed that the compressive power which was applied to the globe, must either force the particles of the liquid into closer adhesion, or drive them through the sides of the vessel before any impression could be made on its surface. Although the latter effect took place, it furnishes no proof of the incompressibility of water, as the Florentines had no method of determining that the alteration of figure in their globe of gold occasioned such a diminution of its internal capacity, as was exactly equal to the quantity of water forced into its pores; but this experiment serves to show the great minuteness of the particles of a liquid in penetrating the pores of gold, which is the densest of all metals.

Mr. Canton brought the question of incompressibility to a more decisive determination. He procured a glass tube, of about two feet long, with a ball at one end, of an inch and a quarter in diameter. Having filled the ball and part of the tube with mercury, and brought it to the heat of 50° of Fahrenheit's thermometer, he marked the place where the mercury stood, and then raised the mercury by heat to the top of the tube, and there sealed the tube hermetically; then upon reducing the mercury to the same degree of heat as before, it stood in the tube $\frac{32}{100}$ of an inch higher than the mark. The same experiment was repeated with water exhausted of air instead of mercury, and the water stood in the tube $\frac{43}{100}$ above the mark. Now, since the weight of the atmosphere on the outside of the ball, without any counterbalance from within, will compress the ball, and equally raise both the mercury and water; it appears that the water expands $\frac{11}{100}$ of an inch more than the mercury, by removing the weight of the atmosphere. From this and other experiments, he infers that water is not only compressible, but elastic, and that it is more capable of compressibility in winter than in summer.

All liquids gravitate or weigh in proportion to their quantity of matter, not only in the open air or in vacuo, but when immersed in similar liquids.

Illustration. Although this law seems so consonant to reason, it has been supposed by ancient naturalists, who were ignorant of the equal and general pressure of all liquids, that the component parts or the particles of the same element did not gravitate or rest on each other; so that the weight of a vessel of water balanced in air, would be entirely lost when the fluid was weighed in its own element. The following experiment seems perfectly to decide this question.

Take a common bottle, corked close, with some shot in the inside to make it sink, and fasten it to the end of a scale-beam; then immerse the bottle in water, and balance the weight in the opposite scale; afterwards open the neck of the bottle, and let it fill with water, which will cause it to sink; then weigh the bottle again. Now it will be found that the weight of the water which is contained in the bottle is equal to the difference of the weights in the scale, when it is balanced in air; which sufficiently shows that the weight of the water is the same in both situations.

As the particles of liquids possess weight as a common property of bodies, it seems reasonable that they should possess the consequent power of gravitation, which belongs to bodies in general. Therefore, supposing that the particles which compose liquids are equal, their gravitation must likewise be equal; so that in the descent of liquids, when the particles are stopped and supported, the gravitation being equal, one particle will not have more propensity than another to change its situation, and after the impelling force has subsided, the particles will remain at absolute rest.

From the gravity of liquids arises their pressure, which is always proportional to the gravity.

Illustration. For if the particles of fluids have equal magnitude and weight, the gravity or pressure must be proportional to the depth, and equal in every horizontal line of fluid: consequently the pressure on the bottom of vessels is equal in every part.

The pressure of fluids upwards is equal to the pressure downwards at any given depth.

Illustration. For, suppose a column of water to consist of any given number of particles acting upon each other in a perpendicular direction, the first particle acts upon the second with its own weight only; and as the second is stationary, or fixed by the surrounding particles, according to the third law of motion, that action and reaction are equal, it is evident that the action or gravity in the first is repelled in an equal degree by the reaction of the second; and in like manner the second acts on the third, with its own gravity added to that of the first, but still the reaction increases in an equivalent degree, and so on throughout the whole depth of the fluid.

The particles of a fluid, at the same depth, press each other equally in all directions.

Illustration. This appears to rise out of the very nature of fluids; for as the particles give way to every impressive force, if the pressure amongst themselves should be unequal, the fluid could never be at rest, which is contrary to experience: therefore we conclude that the particles press each other equally, which keeps them in their own places. This principle applies to the whole of a fluid, as well as a part.

For if four or five glass tubes of different forms, be immersed in water, when the corks in the ends are taken out, the water will flow through the various windings of the different tubes, and rise in all of them to the same height as it stands in the straight tube. Therefore the drops of fluids must be equally pressed in all directions during their ascent through the various angles of the tube, otherwise the fluid could not rise to the same height in them all.

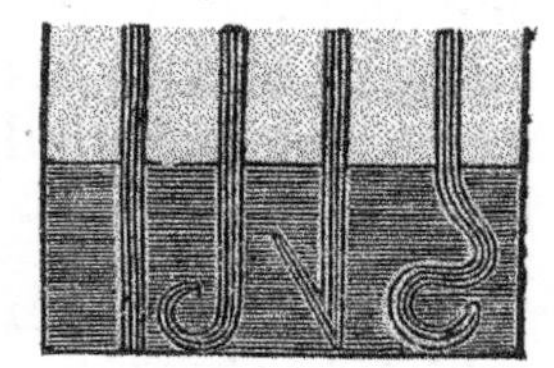

From the mutual pressure and equal action of the particles of fluids, the surface will be perfectly smooth and parallel to the horizon.

Illustration. If, from any exterior cause, the surface of water has some parts higher than the rest, these will sink down by the natural force of their own gravitation, and diffuse themselves into an even surface.

Since fluids press equally every way, the pressure sides of a vessel will be proportional to the depth of the particle from the surface of the fluid.

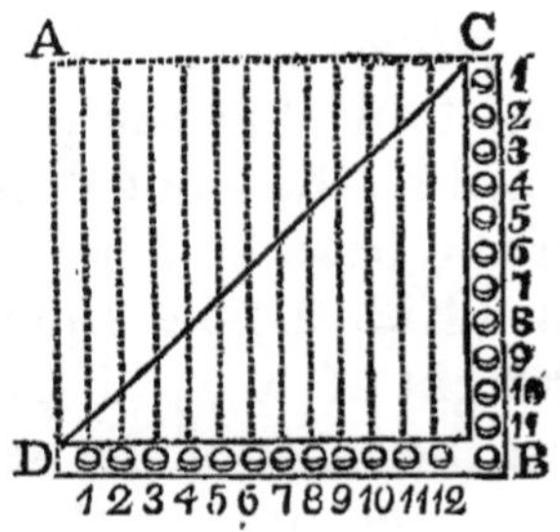

Illustration. For considering the first particle in the line C B to have no other weight upon it, there is no other pressure on the side of the square than that which arises from the particle itself; but the pressure of the second particle is increased by the weight of the first, and the third by the weight of the first and second; thus the pressure keeps augmenting arithmetically to the end of the series. Therefore, considering the pressure on the side as a series in arithmetical progression, beginning with (0), it is equal to half the pressure on the bottom; for as every particle on the bottom sustains 11 others, its pressure will be $11 \times 12 = 132$; but the sum of the series for the pressure on the side of the square is 66, which is equal to half 132. So that the pressure against the side is in proportion to the depth from the the surface, and the whole pressure on the base of the square is equal to the sum of the pressures on both sides. If the figure be made to represent a cubical vessel, the sum of the pressures against the four sides would be twice the pressure on the bottom; consequently the whole force of the fluid would be three times the force of its gravity.

Thus we perceive the difference between fluids and solids. The latter act solely by their gravity; but fluids are not only governed by gravity, but by pressure likewise. Solids act in the perpendicular line of gravitation: fluids press equally in every direction. The force of solids is in proportion to the quantity of matter; but the force of fluids is in proportion to the quantity and altitude.

The pressure which the bottom of a vessel sustains from the fluid contained in it, under every form of the vessel, is equal to the weight of a column of the fluid, the base of which is equal to the area of the bottom, and the height the same with the perpendicular height of the fluid.

Illustration. From what has already been said, we find that the pressure amongst the particles of a fluid at the same depth is every where equal. Then, if the base and height of the vessel A D, (Fig. 2.) be equal to the base and height of A F, (Fig. 1.) the pressure on the base of the second is equal to the pressure on the base of the first, although the capacity or quantity contained in the first is considerably greater than that which is contained in the second. Make I K L M, Fig. 1, equal to E G F H, Fig. 2; then the pressure on the bases L M and F H is evidently equal, and if I M be parallel to C D, there will be the same pressure upwards on every part of the line I M; and the pressure at I, where the column of water touches the side of the vessel, is equal to the pressure of any other column in the line I M; therefore a column extended from I to L would be equal to N E, and the pressure on the side of the vessel at I, from the motion of fluids in every direction, has a reaction equal to the weight of the column N E, which makes the pressure of I K on the base equal the whole column E F. This principle will apply to any other lateral point in C E; so that the whole, taken together, makes the pressure on the bottom C D equal to the pressure of the fluid on the bottom of the cylindrical vessel E F. It may be shown that the lateral reaction at I is equal to the weight of the column N E, by inserting a glass tube in the side of the vessel; for the pressure upwards at I will fill up the tube to the surface of the vessel.

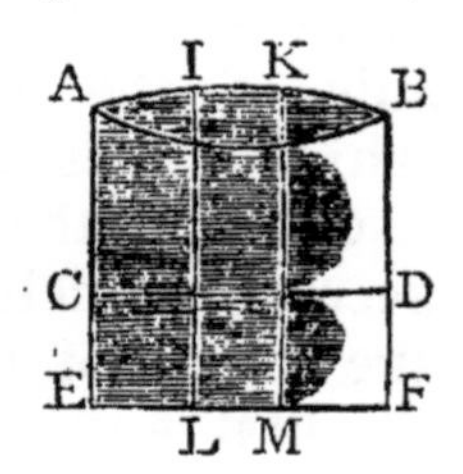

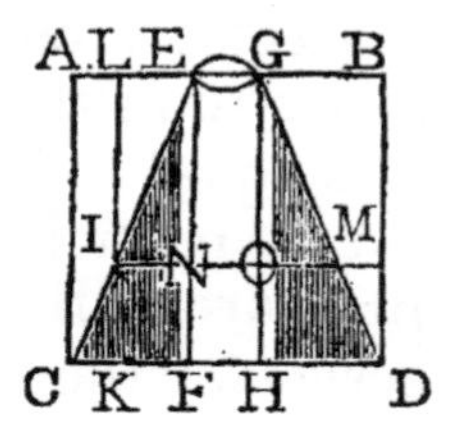

From this it appears, that the pressure on the bottom of a vessel, of what form soever, is not according to the quantity of fluid contained, but according to the perpendicular height.

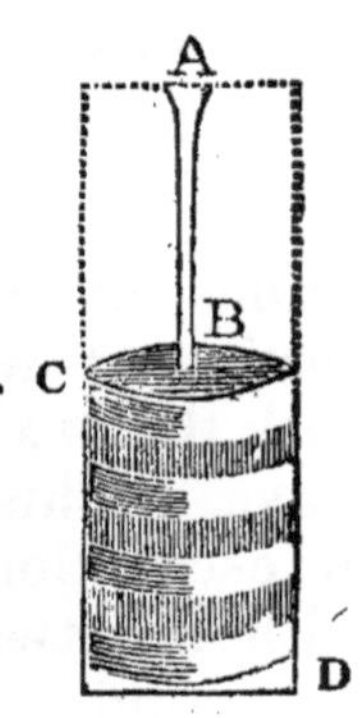

If C D, a hogshead full of water, be placed on its end, and the brass tube A B inserted in the top; on filling the tube with water, the compressive force on the sides of the vessel, consequently the danger of bursting, would be as great as if the whole column were carried up to the top of the tube. As the bottom of vessels supports a pressure proportional to the height of the fluid, so the sides near the bottom have the greatest force acting against them, which decreases in horizontal lines to the surface.

What is called the Hydrostatic Bellows, is a machine well calculated to show that the pressure of fluids arises more from the altitude than from the quantity contained.

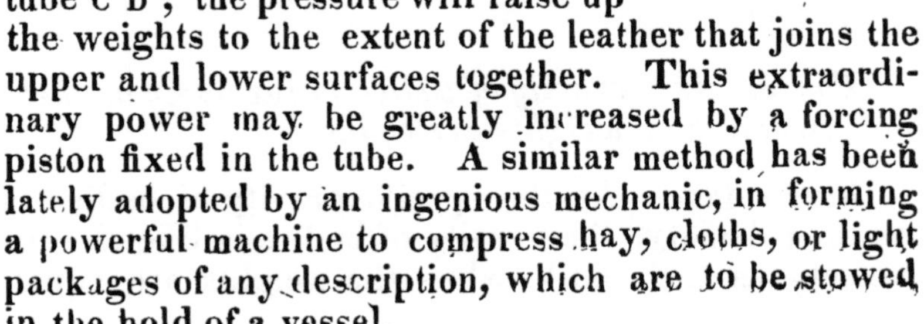

This is formed of two thick boards A and B, about 18 inches long and 16 inches wide; these are joined with strong leather in the manner of bellows, and to the under plank B is fastened a small brass tube C D, which communicates with the interior of the machine. When the experiment is made, a small quantity of water is first poured in to keep the top and bottom asunder; then if 300 weight be placed on A, and the machine be filled with water, till it stands about three feet high in the tube C D; the pressure will raise up the weights to the extent of the leather that joins the upper and lower surfaces together. This extraordinary power may be greatly increased by a forcing piston fixed in the tube. A similar method has been lately adopted by an ingenious mechanic, in forming a powerful machine to compress hay, cloths, or light packages of any description, which are to be stowed in the hold of a vessel.

Liquids may be made to oscillate, on the principle of the pendulum.

Illustration. The waves of the sea perform oscillations, by the alternating ascent and descent of that portion of water which is in the trough of the sea, with the next portion which ascends in the form of a wave. This may be well exhibited by the vibrations or oscillations of water or mercury in a bent tube, with the two open ends upwards, partly filled with the liquid.

Remarks. The density of a body is the quantity of matter contained under a given bulk or magnitude, which is relative as the quantity of matter is to its magnitude; for the greater the number of particles which are contained in a given portion of space, the greater is the density of the body; and the fewer the number contained in a like space, the less is the density.

The specific gravity of a body, is its weight compared with any other body of the same bulk or magnitude. Thus, the specific gravity of lead to water is as 11 to 1; that is, a cubic inch of lead is as heavy as ten cubical inches of water.

The specific gravity of bodies is as their density.

Illustration. For as the specific gravity is the weight of a given magnitude, and as the weight of bodies is according to the given quantities of matter, the specific gravity is as the quantity of matter contained in a given magnitude, or as the density of that magnitude.

The specific gravity of bodies is inversely as their bulk, when their weights are equal.

Illustration. As the specific gravity of bodies is as their density, the density of bodies is likewise inversely as their bulk when the weights are equal.

The specific gravity of gold to lead being as 19 to 11, a bar of gold an inch square and 11 inches in height, will possess the same weight as a bar of lead 19 inches high, and of the same base. The magnitude or bulk of a body is expressed by a number, denoting its relation to some standard generally used, as a cubical inch, foot, &c. The expression of the absolute gravity of a body is likewise relative, being determined by some arbitrary weight, as an ounce, pound, &c.

Remark. The specific gravity of platina compared with water, is as 22 to 1; diamond, $3\frac{1}{2}$ to 1; sound oak, 1 to 1. Dry pine is half as heavy as water; cork is one-fourth as heavy; atmospheric air about the 800th part as heavy.

When a body is immersed in a fluid, it loses just as much of its weight as is equal to the weight of an equal bulk of the fluid.

Illustration. When a solid enters a fluid specifically lighter than itself, it displaces as many particles as are equal to its own magnitude, and these particles oppose its descent with a force equal to their pressure upwards, in a column of which the base is equal to the bulk of the solid; therefore the weight of the solid in the water must be diminished by the weight of the pressure of the fluid. But inasmuch as the gravity of the solid exceeds the pressure of the olumn of fluid upwards, it descends by the excess, losing a part of its weight equal to the repulsive force of the fluid.

This principle may be more clearly understood by the following experiment.

Let E (see next figure) be one end of a scale-beam, and B a bucket made to contain a quantity of fluid exactly equal to the magnitude of the cylinder A, which is fastened to the bucket and scale-beam.

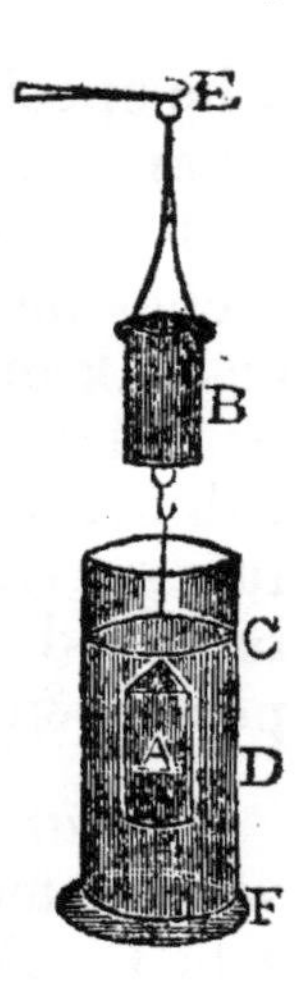

After balancing the scale at the opposite end, immerse the cylinder A in the vessel of water C F; then the end with the weights will overbalance the opposite end with the cylinder: but when the bucket B is filled with a quantity of water equal to the magnitude of the cylinder A, the balance will become equipoised. This evidently shows that the cylinder loses as much of its own weight as is equal to the weight of its magnitude of the fluid; for on adding this weight, the inequality ceases, and the balance is restored.

The weight of a fluid is increased as the weight of the body is decreased.

Illustration. The action or pressure on the bottom of the vessel is augmented in proportion to the bulk of the solid; and as the gravitation of fluids is according to their heights, the power of the fluid is increased by the difference of the height before and after the immersion of the body: for if the immersion of A raise the water from D to C, the accumulated weight on the bottom of the vessel will be equal to the area of D C, which is equal to the magnitude of the body.

If any body lighter than an equal bulk of the fluid, be placed on its surface, it will sink, or descend in it, till it has removed or displaced as much of the fluid as is equal to the weight of the body.

Illustration. When a solid which is specifically lighter than a fluid, is placed on its surface, it sinks till the pressure upwards is equal to the pressure downwards; then the respective powers are in equilibrio, and the weight of the fluid displaced is equal to the whole weight of the solid. Let a globular piece of wood, the specific gravity of which is less than that of the water, be set afloat in a vessel of water; then take the exact weight of the whole, and observe the point to which the water rises by the immersion of the wood. Now, if the wood bé taken out, and the vessel filled up to this point, on weighing the vessel again, it will be found to have the same weight as when the wood was immersed, which shows that the weight of the water which is displaced is equal to the weight of the whole body; therefore the whole solid is to the part immersed, as the specific gravity of the fluid is to that of the solid.

All solids of equal magnitude, though of different specific gravities, lose an equal weight when they are immersed in the same fluid.

Illustration. For as the weight that all bodies lose in water, is according to the quantity of water displaced, the same bulk will displace the same quantity; consequently it will lose an equal weight. Thus, a piece of brass loses as much weight in water as a piece of gold of the same magnitude, although the specific gravity of the gold is twice as much as that of the brass.

Bodies that have the same weight, but different specific gravities, lose unequal parts of their weights when they are placed in the same fluid.

Illustration. If a piece of gold and a piece of brass be balanced in opposite scales, and afterwards weighed in water, it will be found that the gold overweighs the brass: for when their weights are equal, their magnitudes are as their specific gravities; now as the specific gravity of the gold is more than twice that of the brass, the bulk of the gold is much less than the bulk of the brass of the same weight; therefore the bulk of the gold displaces less water, and loses less of its own weight.

If these two bodies be first balanced in water, and then in air, the brass in this case will overweigh the gold: for each of them loses a part of their weight proportionate to their bulk; and as the bulk of the brass is greater than that of the gold, it loses more weight in the water; but this difference is restored when the bodies are weighed in air, which causes the brass to preponderate.

If a solid which is equal in weight to an equal bulk of fluid, be immersed therein, it will remain indifferently in any part of the fluid.

Illustration. For as bodies descend by their own gravity, if a column of water of equal gravity and power be opposed in the descent of the body, the forces will destroy each other, or become equal, so that the body which is immersed will remain suspended in any part of the fluid.

If a body which is heavier than an equal bulk of the fluid, be immersed in it, it will descend by the excess of its gravity above that of the fluid.

Illustration. When a body is immersed, it loses a portion of its weight from the resistance of the medium, and descends by the excess; that is, if the gravity of descending body be equal to 3, and the resistance of the medium equal to 2, the excess or power of descent will be 1.

This relative gravity of solids by which they sink or swim, is amusingly shown by the ascent and descent of glass images in a jar of water.

The images are made nearly of the same specific gravity with the water, but rather lighter, with their weights a little varied, to make them take different situations in the vessel. As the bodies of the images are hollow, they contain a quantity of air, and when they are immersed in the fluid A B, the air communicates with the water by means of a small hole in the heel of each image. The top of the vessel C is covered with a bladder, which includes a quantity of air in the upper part A D; on pressing the bladder, the elasticity of the air presses on the water, and causes it to compress the air in the bodies of the images, which suffers a small portion of water to enter the heel, and increase their gravities; this causes the images to descend: and when the pressure of the air is relaxed on the surface, by taking up the hand, the air which is contained in the images forces the water out of their bodies, and they rise in the vessel. Thus, by varying the pressure, the images may be made to ascend and descend at pleasure.

It is by the pressure of fluids upwards, that bodies, which are specifically lighter than water, rise in it.

Illustration. If any solid body lighter than water be immersed to a certain depth E, the pressure upon the water underneath the body is equal to the body D E added to the column of water A D, which extends from the top of the body to the surface of the fluid: but the ascending pressure of the column of water C E is equal to a column of water A E; therefore inasmuch as the body is specifically lighter than the fluid, so is the pressure E C greater than E A; consequently this superiority of pressure forces the body upwards to the surface of the fluid.

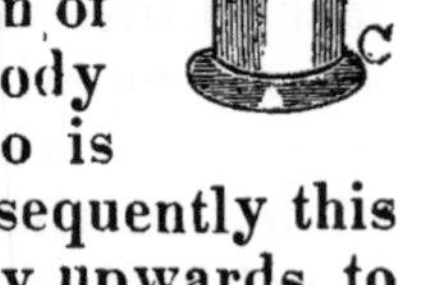

By the reverse of this principle, bodies specifically heavier than water sink to the bottom: for if the body be of greater specific gravity than the water, the pressure of the column A E is greater than the power of resistance E C; therefore the body descends.

Bodies which are lighter than water will not rise in it, if the pressure of the water underneath the body can be taken off.

Illustration. If a smooth and even plate be fitted on the end of the wooden cylinder A C, and placed exactly on another plate B, at the bottom of a vessel of water, the cylinder and plate A C, though specifically lighter than the water, will not ascend; for the action or pressure of the fluid upwards being taken away, the body is kept down, not

only by its own natural gravity, but by the weight of a column of water of the same base with the plate, from the top of the body to the surface of the fluid.

Bodies specifically heavier than water may be made to swim on it, when the water is kept from the upper surface, till the descent of the body meets with a column of fluid, the pressure of which is equal to the specific gravity of the body.

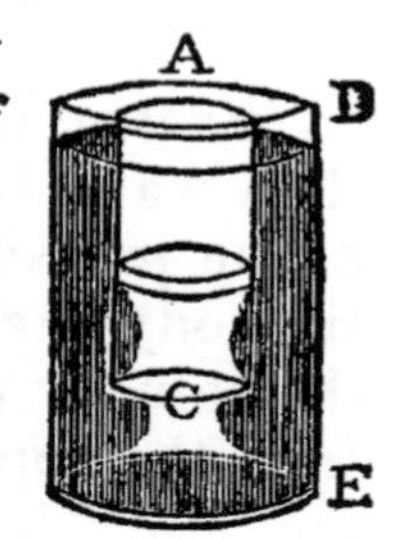

Illustration. As the specific gravity of water to brass is as 1 to 9, if a plate of brass be fitted into an open glass cylinder A C, which is suspended in a vessel of water D E, so that no part of the fluid can get on the upper surface of the brass, when it is sunk nine times its own thickness, it will remain floating on the surface of the water in the cylinder, pressed upwards by a force equal to the weight of a column of water, the height of which is nine times the thickness of the plate, so that it remains supported by a resistance equal to its pressure.

Remarks. As specific gravity implies the relation of bodies to one another, some standard or given quantity must be adopted, by which these relations may be determined; and for the sake of experiment, it is necessary that this standard should be fluid. For this reason, as well as for conveniency, water is used. It is likewise found that a cubical foot of distilled water weighs one thousand ounces avoirdupois, which may be taken as a thousand or as a unity, to show the comparative relation, or the specific gravity of bodies.

To make this subject more clear, it may be necessary to repeat, what we have before stated, that "when any body is immersed in a fluid, it loses just as much of its weight as is equal to the weight of an equal bulk of the fluid; but the weight lost by the body is gained by the fluid, which will be increased in its weight by as much as the body has lost."

According to this principle, we shall have three terms given to find a fourth; that is, the weight of a body in air, the difference of its weight in air and in water, and the given specific gravity of the water, to find the comparative relation of the body.

Suppose a piece of gold weighs 38 grains in air, but when it is balanced in water it weighs only 36, then it loses two grains by immersion, which is equal to the weight of the water displaced by the gold. Now by proportion, as the weight of the displaced fluid is to the weight of the gold in air, so is the given number to the specific gravity of the gold; or,

As 2 : 38 :: 1000 : 19000, or 1 to 19.

That is, if the specific gravity of the water be 1000 or 1, the specific gravity of gold will be 19000 or 19; or, in other terms, gold is nineteen times heavier than water.

By this means, with the assistance of the hydrostatic balance, the specific gravity of bodies in general may be determined.

The hydrostatic scales are of various constructions, according to the accuracy which is required in the experiment; but it will be sufficient for our purpose to describe those that are the least complex, yet sufficiently accurate for common experiments. They are made nearly like common scales, but with much greater nicety, and the strings to the scale at one end of the beam are shortened, so as to admit the water cylinder and body underneath it. When the experiment is to be performed, the body is fastened by a horse-hair to the hook under the shorter scale, and then balanced in air; it is afterwards weighed in the water cylinder, and the difference of weight of the body in the two mediums, shows the difference of the specific gravities, which is equal in weight to a bulk of water of the same magnitude as the body immersed. The relative proportion is then found in figures, by stating the question as we have already shown.

By an improved balance of this kind, the different qualities of gold, or of any other metal, may be ascertained with considerable exactness; for, as all bodies weigh in proportion to the gravitating matter which they possess under the same bulk, and as the specific gravity of fine gold is greater than that of any other metal, except platina, it will possess a greater weight under the same magnitude. In determining the quality of gold by the balance, it is necessary to fix on some criterion for its purity; suppose it to be the common standard gold. First find the specific gravity of this standard by the preceding method; then, by taking the specific gravity of any alloyed quantity of gold, the difference between them will show the quantity of alloy. Thus, suppose a new standard guinea weighs 129 grains in air, and when it is weighed in water, it requires 7¼ grains in the water scale to balance it, its weight being only 122¾ grains in the denser medium; it is evident, from what has been said of bulk and gravity, that any inferior alloy would require a greater weight in the water scale to restore an equal balance.

If a suspected guinea should weigh 129 grains in air, but on trying it in water it requires 8¼ grains in the water scale to produce an equilibrium, it shows the gold to be inferior to the standard, which only took 7¼ grains: thus the difference of purity may be known in every kind of metal, by the difference of gravity from the standard of that metal.

The specific gravity of those bodies which are lighter than water, may be determined in the following manner: Weigh the substance in air, then fasten it to the bottom of the water scale with a stiff wire, and then take the weights out of the opposite scale, allowing for the weight of the wire; afterwards immerse the body in the water cylinder, adding weights to the water scale till the balance is equal; then add the weights of the two scales together, and say, as the sum of the weights is to its weight in air, so is the specific gravity of the water to that of the body.

For as light bodies do not rise in water by reason of their own levity, but from the superior density of the body in which they are placed, the difference of these gravities will be according to their difference of weight.

Suppose a piece of wood weighs in common air 59.5 grains, and when it is fastened to the water scale and immersed, it requires 16.7 grains in the water scale to balance it; then add 59.5 to 16.7=76.2, and say, as 76.2 : 59.5 : : 1000 : 781, the specific gravity of the wood.

To find the specific gravity of fluids is only to find their different degrees of density, which may be done by fastening a weight to the water scale, and afterwards immersing it in the different fluids, noting the weight of the body in each, and the difference of the weights will be the comparative gravity of each.

To find the specific gravity of a fluid in relation to water, (suppose brandy.)

Let a solid be fastened to the water scale and weighed in air; suppose the weight to be 1464 grains, but on weighing it in water, it loses 445 grains, so that the balance weight for this fluid must be 1464—445 =1019. Now place 1019 grains in the weight scale, and immerse the other end in the brandy, and the body will descend, requiring 38.2 grains at the opposite end to restore the equilibrium; then say, as 445 : 38.2 : : 1000 : 86, which, taken from 1000, leaves 914 for the relative gravity of the spirit to the water; so that an equal quantity of the brandy is about $\frac{1}{12}$ lighter than water.

An instrument called *Hydrometer*, is used by brewers and distillers to determine the strength of their liquors.

The neck A B is a piece of brass, or any other metal which is graduated, to show the different depths to which the instrument descends in different gravities of fluids. B is a brass bulb to which the neck is fastened; and C is a weight which is sometimes hung from the bottom to keep the instrument in an erect position when the bulb is immersed in the fluid; and at A is a small shoulder to receive the weights which are laid on the instrument, to adjust it to any particular depth on the graduated neck.

Now, as the resistance of fluids is according to their density, it is obvious that the instrument will sink deepest in those fluids that are the lightest, and this variation is shown by the scale or neck. When the instrument is immersed, the fluid which is displaced by it, is equal in bulk to that part of the instrument which is covered by the water, and in weight to the whole instrument. Then, supposing its weight to be 4000 grains, the different bulks of fluids containing the weight of 4000 grains may be compared, so that if a difference of $\frac{1}{10}$ of an inch take place in the neck by immersing it in two different fluids, it shows that the same weight of the liquors differs in bulk by the magnitude of $\frac{1}{10}$ of an inch of the stem of the instrument.

The specific gravity of fluids may be found by putting an ounce, or any other weight, of distilled water into a glass phial, and marking the height; then empty the bottle, and fill it up to the same height exactly with any other fluid, and weigh them both in a nice balance; the difference of these weights will be the difference of their specific gravities, for their bulks are equal.

TABLE OF SPECIFIC GRAVITIES,

Supposing Rain Water 1000.

Substance	Specific Gravity	Substance	Specific Gravity
Refined Gold	19,640	Oats	472
Refined Silver	11,091	Dry Pease	807
Lead	10,130	Barley	658
Copper	9,000	Crude Mercury	13,593
Iron	7,645	Mercury distilled 511 times	14,110
Tin	7,550	Alum	1,714
Copper Ore	3,775	Nitre	1,900
Lead Ore	6,800	Myrrh	1,250
Adamant, or Diamond	3,400	Verdigris	1,714
Carnelian	2,568	Opium	1,365
Lapis Lazuli	3,054	Beeswax	960
Lapis Calaminaris	5,000	Pitch	1,190
Common Glass	2,620	Honey	1,450
Chalk	2,370	Resin	1,100
Common Sea Coal	1,272	Human Blood	1,126
Ivory	1,826	Distilled Water	993
Boxwood	1,030	Spring Water	999
Oak	925	Sea Water	1,030
Elm	600	Aquafortis	1,300
Ash	734	Oil of Vitriol	1,700
Fir	546	Oil of Turpentine	874
Cork	240	Rectified Spirit of Wine	840
Wheat	757	Burgundy Wine	953

Hydrodynimics.

Remarks. Hydrodynimics is that branch of the philosophy of liquids, which treats of water power when in motion. But this power cannot be exhibited without suitable instruments for making applications of it. The construction and application of these instruments or machines, is denominated the science of Hydraulics. The distinction, therefore, between Hydrodynimics and Hydraulics is little more than nominal.

Hydraulics differs from Hydrostatics in the following respects:

Hydrostatics show the weight or pressure of liquids upon solids, or the particles of a liquid upon one another when they remain at rest: but Hydraulics treat of the application of the power of liquids when they are in motion; as of the force and construction of engines, pumps, mills, fountains, and every other description of water-power machines.

The general effects of liquids are considered as proceeding from the following causes, viz. their own natural gravity or pressure, the spring of compressed air, or the compression of bodies on their surface.

Liquids always tend to form a level, which is denominated seeking an equilibrium.

Illustration. The most natural motion of liquids arises from their own gravity, which always causes them to attain a horizontal position when the course is left open; for a liquid will rise to the same height whether it passes through the regularly curved conduit A B, or through the various turnings of C D, when the intermediate heights are below the level of its extremities.

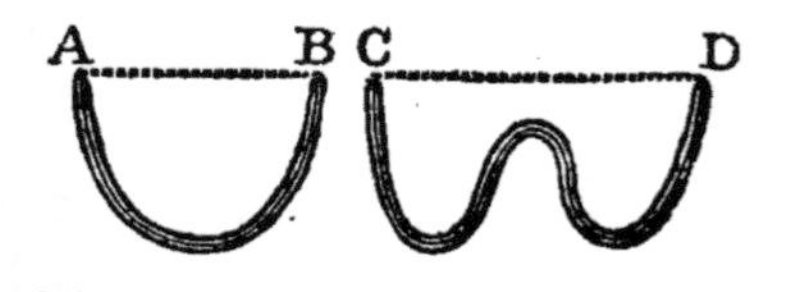

Liquids will flow in air-tight tubes over interposed elevations, if the reservoir is any higher than the place of its final discharge.

Illustration. To supply a reservoir with water at B, from A, the source of the spring; it may be conducted between the two places by pipes laid on the surface of the earth; and notwithstanding the obstruction of the intervening hill, it will flow into the cistern with a velocity equal to that which it would have attained if it had been conducted through the more direct course A D B; for the velocity of fluids is uniformly as their height, which is here represented by the dotted line A C.

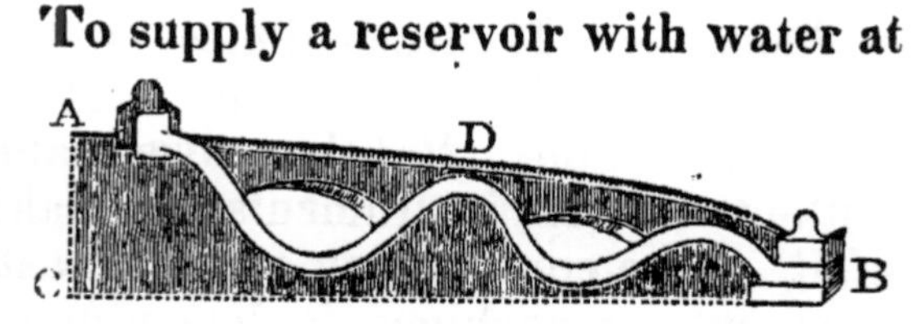

This modern mode of conveying water by pipes is a great saving both in time and expenditure, when it is compared to those stupendous aqueducts which were constructed by the Romans; for in that æra, either from their ignorance of the pressure of fluids, or from their love of magnificence, they conducted water across hills and vallies by straight-lined ducts, which were supported by immense arches or columns.

Water flows with the greatest velocity near the middle and top of the channel of a river.

Illustration. Water is impeded in its progress more or less by friction of the banks and bed of a river. The middle of the channel, or thread, of the river near its top, being the farthest removed from the bottom and banks, is less impeded in its progress. Those who navigate tide rivers, go near the shore to take the earliest advantage of the favorable tide. The momentum being greatest in the middle of the river, the returning tide is resisted longer there than near the shore.

The *Siphon* is a bent pipe or tube, which is used for emptying vessels of liquids, and sometimes for conveying water from one place to another, over hills or obstacles that are higher than the surface of the liquid.

If the small bent tube E F, the legs of which are of equal length, be filled with water and turned downwards, with the ends suspended horizontally, the liquid will not run out; for the gravitating power of the water is equal in each leg, and the upper pressure of the atmosphere is kept off by the form of the machine, which causes the opposing resistance of the air that presses on the surface of the water in the extremities of the pipe to prevent its descent. But as the weight of a column of the atmosphere is equal to a column of water about 34 feet high of the same base, and to a column of mercury 29 inches high in a medium state of the air, if the inverted legs of the siphon exceed these measures for the respective liquids, the gravitating weight will overcome the resistance of the atmosphere, and the liquid will run out.

If the legs are of unequal length, like those in the siphon G H, and the shorter leg be immersed in a vessel of liquid, on sucking the longer end with the mouth to produce a vacuum, or by inverting the tube full of water, the liquid will run out of the vessel till it reaches the bottom of the shorter leg; for the orifice H of the longer leg is exposed to the pressure of the atmosphere, and as the liquid is supported in the shorter leg by the surrounding liquid in the vessel, it is likewise supported by the pressure of the atmosphere which acts on the liquid in the vessel. Now the atmospherical pressures are equal; but these pressures are counteracted by unequal columns of liquid G I and I H; therefore the shorter column G I is more pressed against H I at the vertex I, than the column I H is pressed against I G; consequently the longer column must give way to the greater pressure, and the liquid will run out of the orifice H.

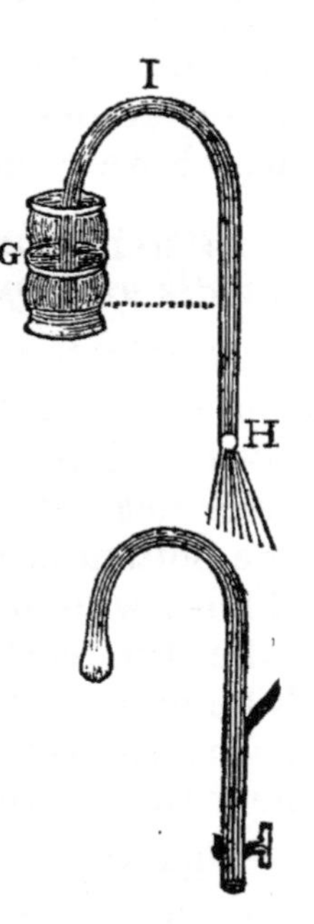

The crane which is used by brewers or distillers for emptying hogsheads, is sometimes made with a cock and small pipe at the end of the longer leg to suck the air out of the tube, and sometimes with a cock only; for when the cock is shut before the shorter end is immersed in the liquid, the air which is pent up in the crane prevents the liquid from rising to the same height as that which surrounds it; but on opening the cock to emit the air, the pressure of the exterior liquid gives such velocity to the interior in rising to the general surface, that it is carried beyond it; and if the curved part of the siphon be not too high above the liquor in the vessel, the interior liquid will fall over it, and continue to flow.

In playing fountains, the force of the jet d'eau is in proportion to the height of the reservoir above the place of discharge.

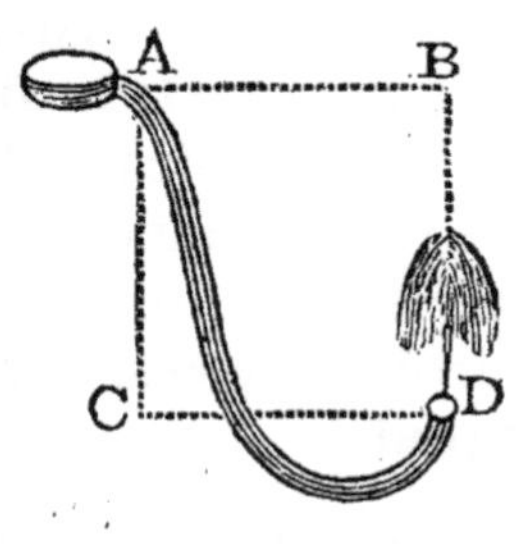

Illustration. Thus the water which descends from a reservoir at A, would acquire such velocity from its gravity as would carry it up to its level at B, if the pipe or tube were continued; but as the pipe terminates at D, it will issue at the adjutage or aperture with a velocity that would have carried it up to B, and equal to that which it has acquired in falling from A to C; so that the velocity of fountains at their adjutage is in proportion to the perpendicular height of their extremities; but the resistance of the air at the lower extremity breaks the column of the fluid and destroys its force, which, joined to friction and other impediments, prevents the liquid from reaching the height of its source.

When a liquid issues from a hole in the bottom or sides of vessels, the velocity of the liquid is equal to that which a body acquires by falling perpendicularly through a space equal to the distance between the surface of the water and the aperture in the vessel.

Illustration. When the height of the liquid is kept up by a constant supply, the velocity will be equal, whatever may be the density of the liquid; for if the pressure be increased by a denser liquid, the issuing quantity will be greater, as velocities are always equal when the moving forces are proportional to the masses which they put in motion. Therefore the quantity of liquid which issues through the same hole in the same time, is in proportion to the celerity of its motion; and as the velocity of bodies in falling through a given space, is according to the squares of the distance fallen through, the velocity of issuing liquids must be according to the square root of their height or pressure.

From the equal pressure of liquids in every direction, the issuing velocity will always be the same at the same depth, whether it proceed from a hole sideways, downwards, or upwards.

The greatest distance to which water spouts from different holes in the side of a vessel, is from that hole which is placed exactly in the middle, between the top and bottom of the vessel; and at the first and third quarters, the projected distances will nearly be equal.

Artificial fountains may be formed by the compression of air on the surface of a liquid.

Illustration. If the vessel B be partly filled with water, and A the upper part of the vessel be filled with compressed air by means of an injecting syringe, the pressure of the air on the surface of the liquid will force it up the pipe, and out of the adjutage, with a force proportional to the power of compression.

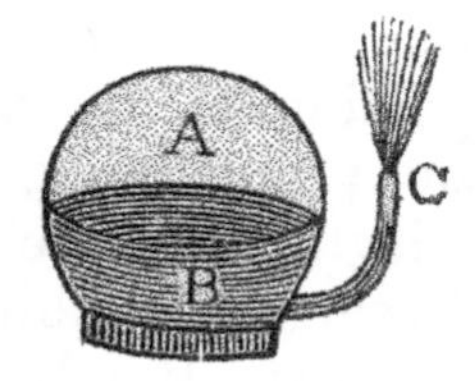

But as this subject has already been explained in Pneumatics, we will only describe a machine which acts by the compression of air, for raising liquors from the cellar to the bar of taverns, &c. A is called the receiving vessel, which is made perfectly air-tight, and sunk about half its depth in the floor of the cellar; the leather hose D is occasionally used to empty butts of liquor E into the receiving barrel, through which it runs by its own natural gravity. After the receiver is filled to a proper height, the communication is stopped between the vessels, and the air is injected into the upper part of the receiver by means of the forcing piston and pipe B, which is placed near the

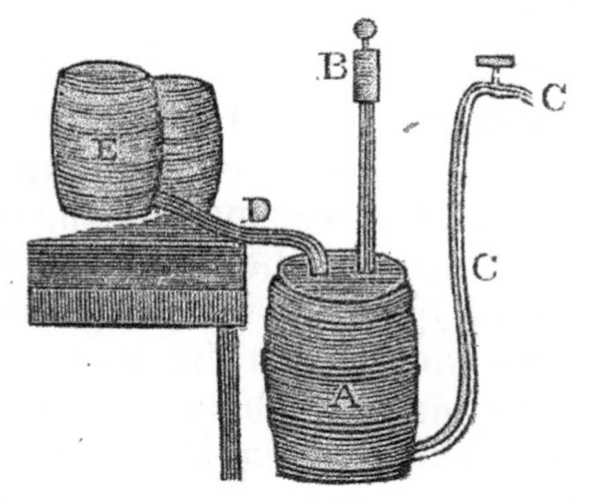

bar. This compressive power on the surface compels the liquor to ascend through C C, which is a leaden pipe with a cock that passes from the bottom of the receiver to some convenient place where the liquor is to be drawn. When the velocity of the fluid decreases at the cock, it may be instantly renewed by three or four strokes with the handle of the piston. The most common machine which is now used for raising beer, is constructed like the following pump.

Remarks. The common *lifting pump* was invented about one hundred and twenty years before the birth of Christ; but it has been greatly improved since the time of Galileo, when the pressure of the atmosphere became more perfectly known.

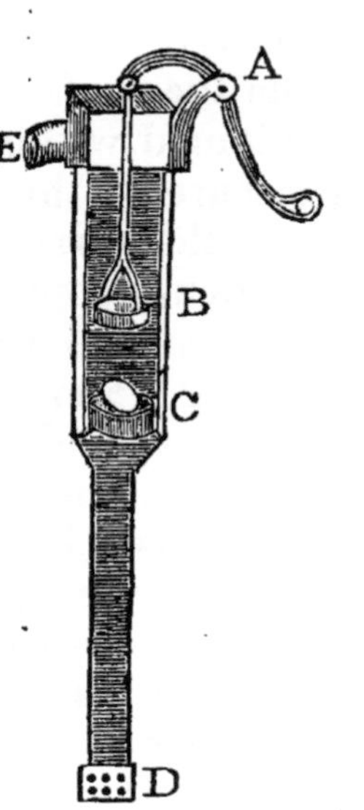

This pump is formed of a long cylinder of wood or lead, one end of which stands in the water at the bottom of the well. It contains two valves, or hollow pieces of wood, which fit close to the cylinder, with lids opening upwards; the lower valve C remains fixed, but the upper valve B is fastened to the piston-rod, and moves up and down by the action of the handle or lever.

The lifting pump raises water by the action of its piston, to any height, from near the top of a column of water which counterbalances a column of the atmosphere of equal diameter.

Illustration. The water in the cylinder of the pump, represented in the preceding figure, stands no higher than the water in the well, and the remainder of the cylinder is empty, or rather occupied by air. Now, when the handle of the pump is raised up, the piston B sinks towards C, which condenses the air between B and C, till its resistance forces open the valve or lid; then the air escapes into the upper and open part of the cylinder. As the piston rises, the air which is contained between B and C becomes rarefied, and the elasticity of that portion of air which is contained in the cylinder, between the lower valve C and the surface of the water in the well, forces open the lower lid, and a part of it escapes into the rarefied space between B and C, which has been formed by the rising of the piston. Thus, by a few strokes of the handle, if the wood or metal of the cylinder be sufficiently close to exclude the air, and the piston and valves be well fitted to the sides of the pipe, the compressive power of the atmosphere will be removed from the surface of that part of the fluid which is contained within the cylinder, and the atmospherical pressure on the general surface of the well will force it up the barrel to any height less than 33 or 34 feet.

Then, supposing the lower valve to be placed at a less distance than 33 feet from the surface, the ascending water will force it open, and get admitted into the cylinder between C and B. When the piston descends, the weight of the water upon the lower valve closes it, and the fluid is forced through the upper by the sinking of the piston; so that when the handle is returned, the water, which now rests on the upper lid, is carried towards the top of the cylinder, and flows out of the spout E; and the supply from the well, by the compression of the atmosphere upon its surface, forces through the valve C into the cylinder, as the upper piston raises the water by the power of the handle.

After the pump has been worked, if the barrel and pistons be good, the water will stand in the cylinder close to the spout, and ready to flow on the first stroke of the handle.

As it is the pressure of the atmosphere alone that forces the water up the barrel of the pump, when the lower valve is more than 33 or 34 feet from the surface of the water in the well, the pressure of the air cannot raise it to the valve, consequently the machine would be useless; but this is prevented by sinking the lower piston in the cylinder till it be actually within the height of the pressure, and by lengthening the piston-rod of the upper in proportion to the depth of the lower: this gives an additional weight of fluid to be lifted each stroke, and the power must be proportionate at the handle. But conveniency requires that this operation should be performed by one person; therefore, to lessen the weight of the column of water which extends from the upper piston-lid to the mouth of the pump, the diameter of the cylinder must be decreased and made proportional to the depth of the well, so that the power may be equal to the operation; but the quantity of water which is raised in an equal time, will be less than when the diameter of the cylinder is greater.

The following table shows the diameters of the barrels, and the quantity which is discharged in a minute at different depths, by the power of a man of ordinary strength; supposing him capable of discharging 27½ gallons of water in a minute, by a pump 30 feet high and four inches in diameter; admitting that the power of the man was increased five times by the length of the lever.

Heights of the Pump above the surface of the Well.	Diameter of the bore where the Piston works.	Quantity of Water discharged in a Minute.	
Feet.	Inches.	Galls.	Pints.
10	6. 93	81	6
15	5. 66	54	4
20	4. 90	40	7
25	4. 38	32	6
30	4.	27	4
35	3. 70	23	3
40	3. 46	20	3
45	3. 27	18	1
50	3. 10	16	3
55	2. 95	14	7
60	2. 84	13	5
65	2. 72	12	4
70	2. 62	11	5
75	2. 53	10	7
80	2. 45	10	2
85	2. 38	9	5
90	2. 31	9	1
95	2. 25	8	5
100	2. 19	8	1

The forcing pump raises water by the lifting process, in conjunction with mechanical compression and the elasticity of compressed air.

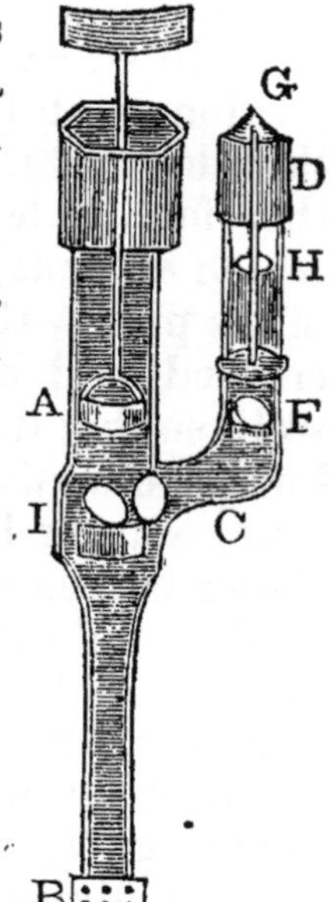

Illustration. This machine differs from the common pump by having a pipe C joined to the barrel, through which the water passes into the air vessel D F; and by the compression of the air which is contained in the upper part H D, the fluid is forced up a pipe, fixed on G, to a considerable height.

By moving the handle, the air is exhausted out of the barrel I B, and forced into the air vessel, and the water follows up the cylinder by atmospherical pressure, in the same manner as in the common pump. But as the piston A is solid, and the closing of the lower valve prevents the water from returning into the well, it is forced through the pipe and valve C F into the air vessel D F, and the valve F closes again by the pressure which rests upon it, whilst the piston ascends to admit a fresh supply into the upper part of the cylinder. The upper part D of the air vessel is made perfectly air tight, and as the water rises in it, it condenses the air by pressing it upwards ; now the air, by its elasticity, reacts on the surface of the fluid in proportion to its density, and forces it up the pipe H G with a velocity proportional to the degree of compression upon the surface ; and as the elasticity of the air makes the pressure perpetual, the pipe produces a continued stream during the rising and falling of the piston.

To gain force, and a perpetual discharge, the air vessel is now used in the construction of engines for extinguishing fire.

The following is the construction of a pump for raising water both by the ascent and descent of the same piston.

The La Hire pump produces a continued stream by compressing the water with both the ascending and descending stroke of the piston.

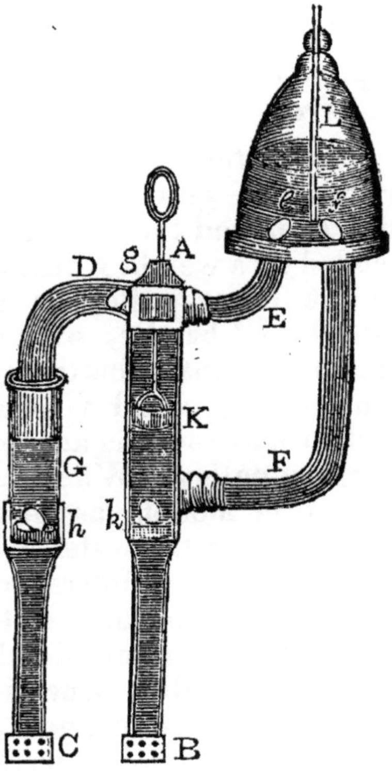

Illustration. The barrel and pipe A B are the same as those of the common forcing pump, and likewise the conveyance pipe F, which carries the water into the air vessel; there is also another conveyance pipe E, which conducts the water into the air vessel that rises in the cylinder D G C, and the piston A K works in a collar of leather at A, which totally excludes the air from the upper part of the cylinder.

When the piston K descends, it shuts the valve *k*, and forces the water up the pipe F into the receiver L, the valve *f* of which immediately closes by the pressure of the fluid. Then, if the air be exhausted out of the adjoining barrel D G C, which communicates with the upper part of the cylinder A K, by the action of the piston, and if the height D C should not exceed 33 feet, it is evident the water will rise up the barrel, and fall through the valve *g* into the cylinder A K as the piston descends; but as the piston ascends, the pressure of the water above it shuts the valve *g*, and forces open another at *e* in the air vessel, by which it enters the receiver; then the return of the stroke which forces the water up the pipe F, closes the valve *e*, and opens *g* again to admit the water into the cylinder. Thus the ascending and descending strokes of the piston force a continual supply of water into the air vessel, whence it is discharged through a conducting pipe by the elasticity and compression of the air, as in the preceding machine. This pump was invented by De La Hire.

The hair-rope pump raises water by passing the ropes quickly through water, and thence to the height required.

Illustration. The three hair ropes F pass in grooves over two pullies A B, and the lines are kept extended by a weight which is fastened to the lower pulley B; at C is a wheel and handle, over which the line passes that joins them to a small multiplying wheel fastened to the well-beam, and this acts on the uppermost pulley. When the machine is put in motion, as the hair ropes pass through the water in the well, it sinks into their interstices, and by the quickness of their motion, it is carried up the ascending ropes in considerable quantities, till it reaches the upper pulley, when it falls into the reservoir E. This method, simple as it may appear, is now used to raise water from a well 90 feet deep; and by tolerable exertion, it is capable of drawing up about 9 gallons a minute.

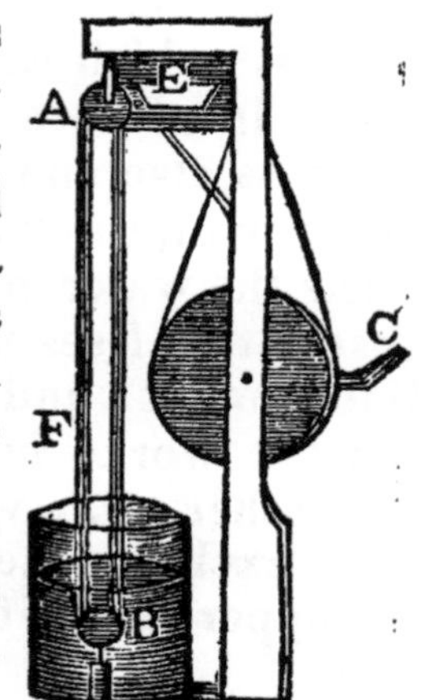

Archimedes' screw-pump raises water by its alternately descending downwards into the lower part of an oblique spiral tube, and being raised to the opposite side by a rotatory motion.

Illustration. This machine, which is now seldom used, is of very ancient date, and is formed by a long cylinder, with a spiral tube from the bottom to the top, through which the water rises till it flows out of the pipe at its upper extremity, as it passes the under side of the cylinder.

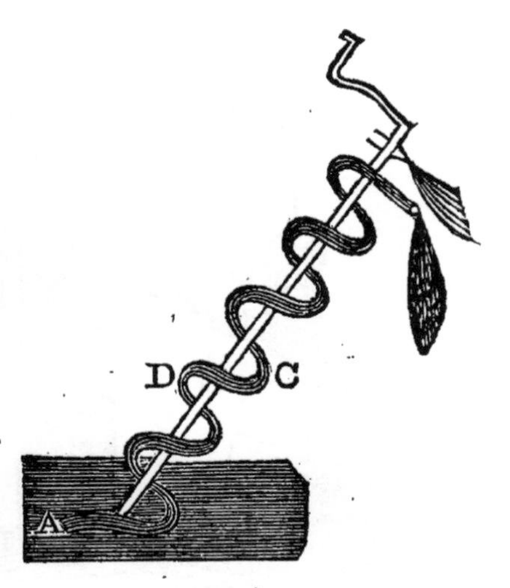

The principle is as follows: When the machine is placed in an oblique direction in the water, the fluid enters at A, the mouth of the spiral, and by the surrounding pressure rises to C; when it has attained this point, it cannot afterwards occupy any other part of the spiral than that which is on the under side; for it cannot move from C towards D, because it is situated higher above the horizon; and as this will always be the same in every similar part, it is evident that when the machine is in motion, the water, as it is raised by the spiral, will always remain on the under side till it flows out of the spout.

Remark. The great improvement, as well as the complexity of the smaller parts of the powerful machine denominated the steam engine, would make a long description of the whole obscure and uninteresting; it will therefore be sufficient for the present purpose, to show the principle of its operation in the most simple state of its construction, as applied to working a pump.

The steam engine is worked by the alternate expansion and condensation of steam.

Illustration. The beam A (see next figure) is placed between two large standards, and turns on its axis B, with a piston and rod fastened to each end, which work in the cylinders D K and G H; the vessel C is partly filled with water, which is kept boiling by a fire underneath it, and this fills the upper part of the boiler with a very strongly compressed elastic steam or vapour. By turning the cock D, the vapour passes through the neck of the vessel, and presses

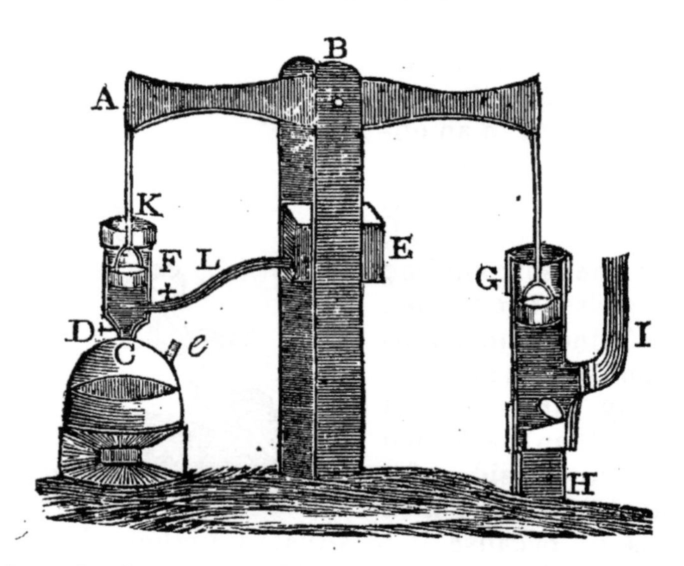

against the bottom of the solid piston F, which forces it up to the top of the cylinder: then the compression which raises the piston F compels the piston G, at the opposite end of the beam, to descend into the cylinder G H, and work the pump, which is of the same construction as the forcing engine that we have already described. The cock D is shut when the piston G is to be raised up to resume its stroke, and the steam in the cylinder K D is instantly condensed by letting in a small quantity of water from the reservoir through the pipe L, which, by destroying the repulsive force of the vapour, suffers the piston F to descend in the cylinder. Methods are in use for condensing steam both above and below the movable piston; but the above description is sufficient for an illustration of the general principles of the machine.

Remarks. When the boilers are so large that the quantity of steam is sufficient to make 20 or 25 strokes in a minute, and each of them 7 or 8 feet high in cylinders 9 inches in diameter, the engine will discharge about 320 hogsheads in an hour.

In the improved construction of steam engines, the operation of turning the cocks is performed by the machine itself, which not only saves the attention of one or two persons, but likewise performs the duty with much more regularity, and causes less danger from the dreadful effects of bursting the boiler when the vapour is not properly discharged. To prevent the bursting of the boiler by an extraordinary expansion of steam, a valve or regulator is made in the side of the vessel, which is forced open by the vapour, to evacuate it when it has acquired a certain force in the boiler.

The steam which is raised by the ordinary heat of boiling water, is about 3000 times as rare as water, and 3½ times as rare as air; and the expansive power of steam against the sides of a globe of copper four inches in diameter, when the water is boiling in it, has been calculated at 38,250*lbs.* weight.

Great improvements have been made in steam engines by Messrs. Boulton and Watt, of Soho, near Birmingham. One of these powerful machines, which was constructed by them, now works a pump 18 inches in diameter, and 600 feet high; the piston makes 10 or 12 strokes, of seven feet long, in a minute, and raises a weight equal to 80,000*lbs.* fifty feet high in the same time, which is performed with a fifth part of the coal that is usually consumed by a common engine.

The present improvements in steam engines fit them for a variety of purposes where great power is required; such as raising water from mines, blowing large bellows to fuse ore, supplying towns with water, grinding corn, &c. Mr. Boulton has lately constructed an apparatus for coining, which moves by an improved steam engine. The machinery is so ingeniously constructed, that four boys of ten or twelve years of age are capable of striking 30,000 guineas in an hour, and the machine itself keeps an accurate account of the number which is struck.

The application of the steam engine to navigation by Messrs. Livingston and Fulton, is now adopted in every part of the civilized world. Several proposed improvements upon Fulton's invention are before the public; but they are not yet sufficiently tested by experience.

ELECTRICITY.

Electric power attracts light substances, when it is excited by friction upon amber, sealing wax, resin, glass, tourmalin, &c.

Illustration. Warm a glass cylinder of any size, and rub it smartly with a clean dry silk handkerchief, and it will attract fibres of tow or flax, down, fine bits of paper, &c. especially if the atmosphere is dry. Almost any amalgam will aid the process, if rubbed on the handkerchief.

Remarks. The attractive power of amber was known some centuries before the Christian æra, but it was then considered as a mere quality which was attached to that peculiar body. But electricity is now supposed to be an elementary power of nature, which is diffused throughout the whole atmosphere, and is attached to bodies in general. Thus, the electrical fluid, which had escaped investigation for many ages, is now become an important branch of science. About the year 1745, we find this subject diffusing widely under the splendid talents of Watson, Canton, and Priestley, in London; Franklin, in America; and the Abbe Nollet, in France. In the hands of these philosophers, electricity has made more progress in a few years than it had gained in all the preceding ages. It was at this time that the mode of accumulating electrical fluid on the surface of glass was carried to its greatest height, by means of what is called the Leyden Phial, from the birthplace of the inventor, who was a native of Leyden. But the greatest discovery that was ever made in electricity, was reserved for Dr. Franklin, in America.

It had been imagined that a similarity existed between lightning and the electrical fluid; but Franklin brought this supposition to the test, and proved the truth of it by the simple means of a boy's kite covered with a silk handkerchief instead of paper, and some wire fastened in the upper part, which served to collect and conduct the fluid. When he had raised this machine into the atmosphere, he drew electrical fluid from the passing clouds, which descended through the string of the kite as a conductor, and was afterwards drawn from an iron key which he tied to the line at a small distance from his hand. By this simple means, he proved that the fluid which produces lightning was exactly the same as that which he obtained from his electrical machine. This important experiment immediately led to the formation of conductors to secure buildings, ships, &c. from the dreadful effects of lightning.

The present opinions on electricity are principally divided into two parts; one relating to what is called *vitreous* and *resinous*, and the other to *positive* and *negative* electricity. The former of these opinions was first laid down by M. du Fay, and was afterwards new-modelled by Symer; but it is now generally rejected. This theory supposes that electrical matter is formed by two distinct fluids, which are repulsive with respect to themselves, but attractive to one another; and the electricities are called vitreous and resinous, from their respective excitability by glass and resin. It is further supposed, that these fluids are attracted by all bodies, and exist in intimate union in their pores, without any exterior mark of existence, until the two fluids be brought into action by a separation of their parts, which is produced by friction. When these electricities are collected and separated by the attrition of the rubber on the surface of the cylinder of an electrical machine, the vitreous passes over to the prime conductor, whilst the resinous is drawn to the rubber. In this state of separation, they exert their respective qualities; so that by electrifying light bodies with each kind of fluid, those that possess the vitreous will repel each other, as well as those that are mutually electrified with the resinous: but if two bodies which are oppositely electrified be brought near together, they will attract each other, and give and receive at the same moment an equal portion of their respective electricities. According to this theory, the electric spark has a double current, and the electrified body will receive from any conductor in the electrical atmosphere an equivalent of the opposite fluid to that which it gives; so that if the finger be presented to the prime conductor of the machine, whilst the body inhales the vitreous stream from the conductor, it gives an equal stream of the resinous from the body: these quantities are so exactly alike, that a light body may be suspended by the opposing forces between the end of the finger and the conductor.

The other theory of *positive* and *negative* electricity, which is now almost universally received, was first taken up by Watson, but was afterwards illustrated, digested, and confirmed by Franklin, and from thence it is called the Franklinian Theory. It supposes that the whole phenomena of electricty depend on a subtile and elastic fluid, entirely of the same kind, repellent amongst its own particles, but attracted by all bodies, and universally disseminated among them. When bodies hold their own natural quantity undisturbed, they are said to be in a non-electrified state; but when the equilibrium is disturbed, that is, when the natural quantity of fluid in a body is varied, either by adding more to that which it naturally possesses, or by taking away a part of its natural quantity, it has an electrical appearance, or it is in an acting state. When a body possesses more fluid than its natural quantity, it is called plus, or positive; and when it contains less than its natural quantity, it is called minus, or negative.

Electricity can be excited with those substances only which are bad conductors, or non-conductors of this power.

Illustration. The progress of this fluid depends

on the nature of the bodies through which it passes. Those which give it the greatest facility in its course, are called conductors, and the fluid is instantaneously transmitted through them even to the greatest distances. Those bodies which will not admit the transmission of electrical fluid, are called electrics, and are impermeable, so that there cannot be an accumulation on one side without a deficiency on the other; and when the two sides are joined together by a proper conductor, the superior quantity, or the positive, rushes through it to the inferior quantity, or negative, till the fluid on both sides of the body be in equilibrio, or in its natural state. When an electric is rubbed by a conductor, as the friction of the rubber upon the glass cylinder of an electrical machine, the fluid is carried from one to the other, and the rubber will be the negative. But as an insulated cushion only affords a small portion of electric fluid, a conducting chain, communicating with the earth, is connected with it, which gives a constant supply from the negative to the positive side. Thus it is conceived, that bodies which are in-different electrical states will readily approach, but that those which are equally charged have an equal repellency.

Electric power, pretty strongly excited, becomes luminous, as well as attractive.

Illustration. If a glass tube, about an inch and a half in diameter, and three feet in length, be rubbed briskly with a piece of leather previously rubbed with an amalgam, in a darkened room, small divergent flames will fly off with a crackling noise, and sometimes a spark of fire six or eight inches long may be seen following the hand upon the surface of the tube. If a brass ball be suspended from the tube by any conducting body, such as a piece of wire or thread, the electric fluid will descend through the conductor and electrify the ball, which will give a spark to the knuckle, or electrify any light body that is presented to it. When the ball is suspended from the tube by a silken string instead of wire or thread, the excitation of the glass will produce no sign of electricity in the ball; and if the down of a feather, or any other light body, be presented to it, it will remain unmoved; for as silk is not a conductor, the fluid cannot pass from the glass to the ball.

Conductors of the electric power will not retain the power an instant, unless they are insulated.

Illustration. Insulated conductors are conducting bodies supported or surrounded by an electric or non-conductor, so that the communication with the earth is cut off. A brass ball and thread suspended from a glass tube, is an insulated conductor; for on excitation, the fluid passes from the non-conducting tube through the thread to the ball, where it is retained by the surrounding atmosphere as an electric, though an imperfect one; or if the ball be suspended by a silken string, and an excited tube be brought to the ball and afterwards taken away, the electrical fluid which is communicated will remain insulated by the air and the non-conducting body of the silk.

The greatest quantity of electricity is obtained by the friction of a conducting body upon the surface of an electric. If the rubber be afterwards insulated, the non-conducting surface will remain charged with the electric fluid, and communicate electric sparks to any conducting body that is presented to it.

If a conducting body be insulated and electrified, the whole of the fluid which is collected will be carried off by a single spark drawn by a conducting body; for as the fluid passes with the greatest facility through all parts of a conductor, the whole flies off at the instant of communication: but non-conductors that are charged, only part with that share of their fluid which lies on the surface next to the conductor.

A mutual attraction exists between electrified and non-electrified bodies, and repulsion between those which are similarly electrified.

Illustration. If a light substance be placed near an electrified body, it will fly towards it till it have obtained the same intensity of fluid, then it will be repelled and attracted by any non-electrified body that is near it. If a non-electrified body be set at a proper distance from an electrified body, and a feather be placed between them, the feather will be alternately attracted and repelled by each; for when it is electrified, it flies to the non-electrified body, and delivers its electricity; it is then attracted and charged again by the other: and thus it will continue its course backwards and forwards, till it have reduced the surplus of fluid in that which is electrified.

If a large globe, cylinder, or plate of glass, be strongly rubbed by a cushion coated with an amalgam, the electric power will exhibit attraction, light, and heat. A machine constructed for this purpose, is called an Electric Machine.

Illustration. As the excitation which is produced by the hand with a rubber on a tube or plate of glass, is not only very laborious, but inadequate to the production of any material quantity of electrical fluid, machines have been constructed of various forms for this purpose; some with spherical glass electrics, some with cylinders, and others by the revolution of a circular plate of glass between cushions or rubbers placed near the edge: but as the cylindrical machine is the most common, and perhaps the most useful, a description of it may be desirable.

In the plate, (see next figure) A represents a glass cylinder called an electric, or non-conducting body, the axis of which is supported by the two sides M M, and these are fixed into the plank K, which is the basis of the machine; C is a common winch or handle by which the cylinder is turned, and D is the cushion or rubber, which is supported and insulated by the glass pillar F. The lower end of F passes into a socket that is acted upon by the screw S, for the purpose of increasing or diminishing the pressure of the cushion

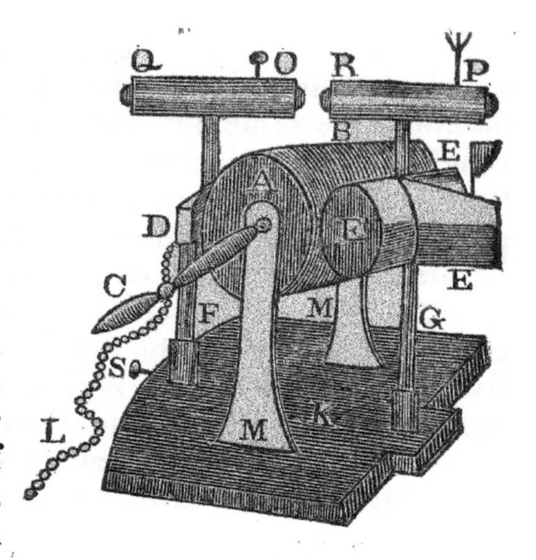

on the surface of the cylinder. B is a piece of black silk which prevents the electrical fluid from flying off, and reaches nearly to the receiving points fixed in the conductor E E; for the closer the silk adheres to the cylinder, the greater will be the degree of excitation. E E is a metallic body, which is called the prime conductor; this is made in various forms, therefore the present T form is not essentially necessary: the receiving points are fastened to the side opposite the cylinder, and the whole is supported from the frame by the insulating pillar of glass G, so that the electrical fluid which is collected on the prime conductor cannot disperse, but remains accumulated for the purpose of experiments. A small quadrant electrometer fixes into a small hole in the conductor, and shows the increase or decrease of the electrical fluid by the rising or falling of the index upon the edge of the quadrant. The chain L has one end fastened to the cushion, and the other lies upon the floor or table, to serve as a conductor for the electrical fluid in passing from the earth to supply the machine; when this chain is taken off, or unconnected with the earth, the machine becomes insulated, and it will retain the electricity that it has acquired during its operation.

In damp weather, it will be of considerable advantage to the power of the machine, to place it in the gentle warmth of the fire for some little time before it is used. When the machine is perfectly dry and free from dust, take an amalgam of zinc and mercury, and spread a little of this composition smooth and even on the cushion, previously rubbed with cold tallow; then apply it to the surface of the cylinder, and turn the handle till the friction becomes tolerably strong, which will give it a great degree of excitation, and prepare it for the general purpose of experiment.

Electric fluid is given off from the positive side in diverging rays, and is received on the negative side in a condensed globular form.

Illustration. If the conductor Q, (see the figure of the machine,) with a small point at O, be placed on a brass rod connected with the cushion, and another with the brass point P be supported by the prime conductor, those bodies which are electrified by Q will not only be attracted by the conductor R, but the electrical fluid will diverge in a *conical* form from the point P, as emitting its electricity, whilst a small, faint, *globular* flame will be seen on the point O, as if it were imbibing the electric stream.

If the glass cylinder be less affected by the friction than the rubber, it will be negatively excited, and the rubber positively. Patterson.

Illustration. The state of positive and negative electricity is governed by the quality of the cylinder and rubber; for if a glass tube be made rough by grinding the surface with emery, and excited by soft flannel, the electricity excited in the tube will be negative; but if it be rubbed by a dried oil-silk and whiting, it becomes positive: even a polished cylinder may be rendered negative by rubbing it with the hairy side of a cat's skin. A cylinder made of baked wood, rubbed with a smooth rubber of oiled silk, becomes negative; but by rubbing it with coarse flannel, it is rendered positive.

Cylinders of sulphur or resin, when rubbed in the usual way, become negatively excited.

Illustration. A cylinder made of sulphur or resin, has the electricity the reverse of that which is produced by the smooth glass cylinder and rubber of the usual machines; for the rubber in this case partakes of the positive, and the cylinder, or the prime conductor, is electrized with the negative.

This difference between the resin and glass gave rise to what is called the double current, or vitreous and resinous electricity; but it is now generally supposed that the difference arises more from the effect of the surfaces that act on each other, than from any peculiar qualities in the different bodies.

Points, attached to any conducting body, either receive or deliver electrical fluid more freely than flat or round bodies.

Illustration. To show the superior attraction of points: If the round knob of a brass conducting rod be held near to the prime conductor when the machine is in motion, the electric spark will be seen darting towards it; but if a needle, or fine pointed conductor, be presented even at twice the distance of the knob, the sparks will instantly cease, and the fluid will be silently drawn off by the point; but when the point is withdrawn, the spark will immediately recommence and fly towards the brass knob. If this experiment be performed in a darkened room, a small globular spark appears at the end of the point when it is presented, which shows that the needle receives the fluid from the conductor. When the wire or needle is fixed towards the end of the prime conductor, on presenting a brass knob, or the finger, the fluid will pass off without any visible appearance; but the electric stream will produce a current like wind, which may be sensibly felt.

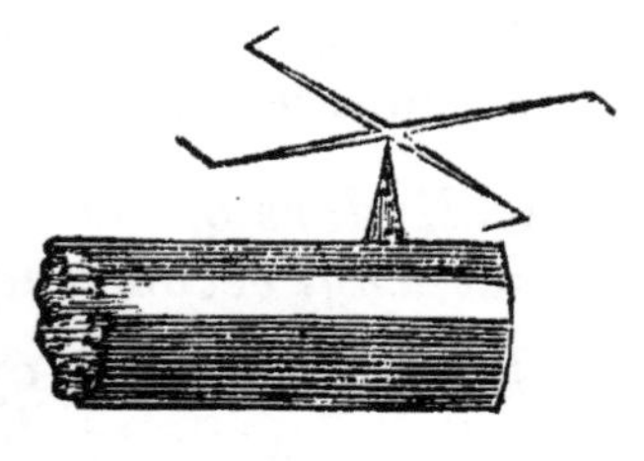

When the needle is fixed perpendicularly on the prime conductor, if crossed wires, with their ends all bent the same way, be balanced on the point, the resistance which the air gives to the electric current that issues from the points, will drive the fly round with considerable velocity.

An amusing experiment, called the electrical orrery, is performed by means of the current which issues from electrified points.

A piece of bent wire is suspended by a needle in the top of a glass stand, and a small globe of glass is fixed in the centre; at one end of the wire is another needle, which supports a short cross wire bent at each end; L is a pith ball placed upon it, which represents the earth; M is a smaller one at the end of the wire for the moon, and S, the small glass globe over the stand, may represent the sun. When the conducting chain is connected with the needle in the top of the stand, and the machine is excited, the sun turns on his axis, and the moon makes her monthly revolution round the earth, whilst the earth is carried in its annual orbit round the sun.

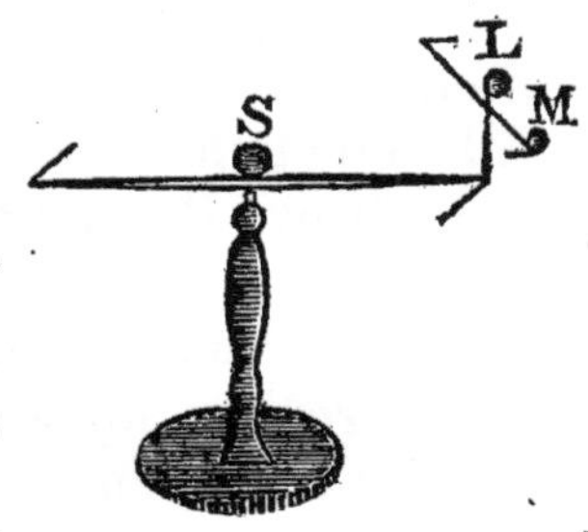

Electricity attracts all kinds of bodies, but repels them after they are electrized.

Illustration. If a piece of light downy feather be suspended at about the distance of a foot from an electrified conductor, it will be attracted or drawn towards it, and afterwards repelled or driven off; for the prime conductor attracts the body till it becomes electrified, after which the conductor and the feather repel each other, and the feather is driven off till it has discharged the power which it had accumulated; then it returns on the same principle as before.

This attracting and repelling power is whimsically illustrated by droll figures cut out in paper. The figures are laid on a metal plate and stand, which is placed exactly under another brass plate suspended by a chain from the prime conductor. When the machine is excited, the upper plate is electrified, and draws the figures towards it; but when the figures are electrified by the upper plate, they are repelled, and fly back to the lower uninsulated stand to discharge their electricity, and then they are attracted again as before. Thus the figures continue jumping backwards and forwards till they have discharged the electric power from the plate and conductor.

The electrified bells is another pleasing experiment, which shows the attraction and repulsion of the electrical power.

Small bells are suspended by small wires from the end of cross rods, and each arm suspends a clapper hung by a silken string; the upper part of the stand is made of solid glass, and the conducting chain of the machine communicates with the brass knob and fly on the top; towards the lower part of the stand is another bell larger than the rest, which is uninsulated, and forms part of a conductor with the earth. When this machine is put in motion, the fluid passes down the conducting wires and electrifies the bells; these attract their respective clappers, which are afterwards repulsed and driven off to the uninsulated bell in the centre, which receives their electricity, and conveys it through the bottom of the stand to the earth. Thus the five bells are kept continually ringing, which produces a pleasant peal when the tones of the bells are properly varied.

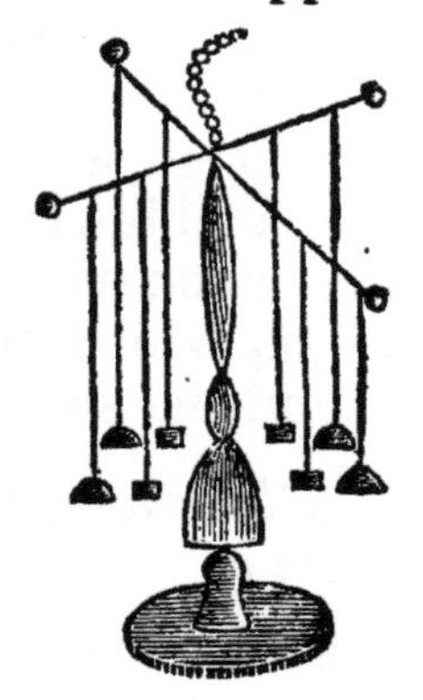

The electric power exhibits its properties strongly by passing from a metallic coat on one side of glass to a coat on the opposite side.

Illustration. What is called the Leyden phial, is a glass jar coated inside and outside with tin-foil, except about two inches on each side from the top of the

jar downwards, to prevent the connexion of the fluid between the inside and outside when the glass is charged. The mouth of the jar is covered by a piece of wood, which receives a thick brass wire; the upper end of the wire has a brass knob fastened to it, and the lower end, which goes into the jar, has a small wire or brass chain fixed to it, that communicates with the bottom and sides, and serves as a conductor to charge the jar with electrical fluid.

When the jar is to be charged, it may be held in the hand, or placed upon a table with its knob against the knob of the conductor; on exciting the machine, the electrical fluid passes from the prime conductor through the knob and wire into the interior of the jar. The fluid thus collected and condensed, will be of the same kind as that which surrounds the prime conductor. The exterior part of the jar being uninsulated, gives its natural electricity to the earth through the medium of the conducting bodies that connect them, and it will require an electricity of the same kind as that which belongs to the rubber; thus the fluid is insulated, with respect to its connexion from the opposite sides of the jar, by the unfoiled part at the top, and the increase in the interior is in proportion to the decrease on the exterior side.

When great force is required from the electric fluid, a number of jars of the above description are placed on a metal frame which forms a communication between their outside coatings and the earth, and the insides of the jars have conducting wires which pass to the prime conductor. In this manner any number of jars may be charged with the same facility as a single one, and from the powerful effect of the electric fluid when it is thus collected, it is called an electric battery.

The bottle form is not absolutely necessary in combining electrical fluid; glass plates were used for the same purpose before this invention was known: indeed it may be combined by bodies of every form, but as cylindrical jars offer the largest surface and greatest conveniency in the smallest space, they are the more usually preferred.

When one side of the jar, or other coated glass, is electrified positively, the opposite side is always electrified negatively.

Illustration. Whatever quantity be thrown on one side, that on the other is diminished in like proportion, so that the charge of an electric jar seems to be the drawing off the fluid from one side and carrying it to the other; for it is impossible to charge one side of a jar, unless the opposite side have a conductor to carry off the fluid which it contains. In like manner, an electrical jar cannot be discharged without a communication between the opposite sides to restore the electrical fluid to its natural quantity. To explain this subject more particularly: If the knob of a coated jar be held near the conductor, on turning the machine, it will be seen by the index of the electrometer when the jar has received its full charge; then take a discharging rod with a glass handle, and bring one of its knobs to the knob of the jar, and the other to the outside coating, which forms a conducting circuit between the inside and outside of the jar, and the surcharge of the interior will fly through the conducting rod to the exterior, till the powers on each side are equal. A person may convince himself of the transmission and force of the fluid through the wires, by forming the conductor himself. For if he touch the coated side of the jar with one finger, and bring the other to the knob of the jar, he will then receive a strong shock, which will be particularly felt, either in his wrists, elbows, or breast, according to the strength of the charge. If the electrical circuit be made by any number of persons joining hands, when the first and last touch the knob and side of the jar, or take hold of two pieces of wire which are joined to them, the whole number will receive the shock at the same instant, be it ever so great; for the passage of electricity is so instantaneous, that in a circuit of wire ten miles in length, the shock was felt at the same moment that the jar was discharged.

In charging a jar, if it be held by the knob, and the coated side presented to the conductor, the exterior will receive the surcharge of the fluid, and the interior will lose it ; that is, the outside will be positive, and the inside negative ; but the effects in the discharge will be the same.

If the jar be placed on an insulated stand, with its knob near the prime conductor, the index will rise by the accumulation of the fluid upon the conductor ; but on trial, it will be found that no electricity has entered the jar. It has been already observed, that the electrical jar is charged by taking the fluid from one side and carrying it to the other ; so that in this case, as the outside has no communication with the earth, it cannot part with its natural electricity : therefore none can be accumulated on either side of the glass.

When the jar is thus insulated, if one of its sides communicate with the rubber by means of a wire, and the other be connected with the prime conductor, the jar will be readily charged with its own fluid ; for the electricity on one side passes through the wire to the rubber, and thence through the prime conductor to the opposite side of the jar. If the knob of the electric jar be placed about half an inch from the end of the prime conductor, and a pointed wire be presented to the coated side of the jar, the fluid will be seen entering the jar from the conductor, and passing from the outside, under the appearance of a small star, upon the point of the wire which is held towards it ; but if a piece of wire be fastened round the jar, with an end projecting towards another conducting wire, the fluid will rush from the side by this point, and diverge its luminous rays in a conical form, taking the point of the wire as the vertex.

Holes may be made in paper, &c. by placing it in the electrical circle, when it passes from a metallic coat on glass to its opposite.

Illustration. Place a card against the tin-foil on the side of the jar; and if one end of the conducting rod be brought against the card, and the other to the knob, the fluid will rush through the conductor, and pierce the card as it joins the other fluid on the exterior surface of the jar.

The surface of glass may be impregnated with gold or silver leaf, by the electric power.

Illustration. Take two slips of glass, about an inch broad and three or four inches long, and place a narrow slip of gold leaf, about an inch longer than the glass plates, between them, so that each end of the leaf may project about half an inch beyond the ends of the glass; then lay the whole upon a non-conducting surface, with a weight upon it, and bring one end of the conducting chain or wire from the bottom of the jar into contact with that end of the leaf which is opposite to it; then bring one knob of the discharging rod to the knob of the jar, and the other to the opposite end of the leaf: thus the electric circuit is completed, and the leaf will be melted and driven into the surface of the glass by the force of the electrical fluid.

A single discharge of the electric power will exhibit numerous sparks, by passing from one metallic patch to another patch on the glass surface.

Illustration. What is called the spotted bottle, is fitted up like the Leyden phial, only the tin-foil coating is gummed on in little square pieces at some distances from each other; so that when the bottle is charged in the dark, the sparks will be seen flying across the spaces, from one square to another. If it be discharged gently, by bringing a pointed wire gra-

dually to the knob of the jar, the fluid will pleasingly illuminate the uncoated parts, and make a crackling noise in passing the spaces.

A double set of bells will show the course of the fluid in a Leyden phial.

Illustration. Let the bottle be placed horizontally in the frame of an insulated stand, with a knob and wire in each end, communicating independently with the inside and outside of the jar,

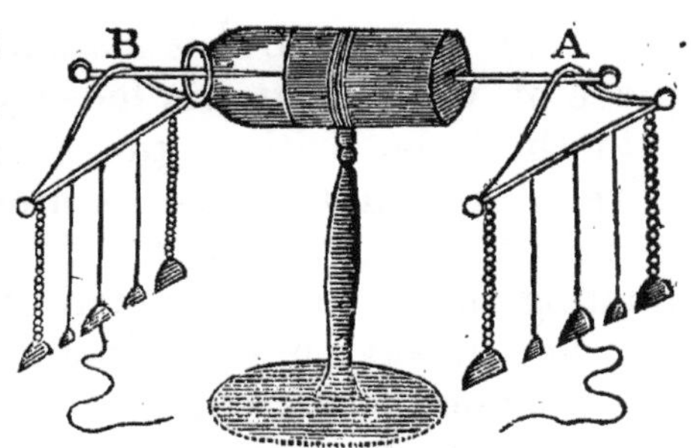

and let a set of bells be suspended from each wire; then place the knob B against the conductor, hang up the chain that lies on the table, and charge the jar. Now, whilst the jar is receiving the fluid, the bells on the opposite wire A, which are connected with the outside of the jar, will continue ringing. After the jar has received its charge, unhook the chain at the end B, and let it lie on the table; then touch the opposite end A with the finger, and those bells at B will begin, and the others will cease: if the finger be again placed on B, those at A will commence. Thus, by varying the end, each set may be rung, till the whole of the fluid be discharged out of the jar.

Buildings may be set on fire by lightning, or the electric power, when combustible bodies are laid near metallic rods, plates of iron, &c.

Illustration. Take some powdered resin, rub and mix it well in dry tow or cotton, and put it between two metallic balls, which are placed in a tin toy resembling a house; from these balls there are two wire conductors, one of which passes to the outside of the charged jar, and the other is connected with one end of the discharging rod. When the jar is discharged, the fluid will instantly force its passage between the balls in the house, and fire the tow or cotton, which blazes out of the windows, and shows some appearance of a house on fire.

A D represents the gable end of a building which is made of wood, with a square hole in the middle; and G B C H is a small piece of wood that fits loosely into the hole, with a piece of wire G H laid through its diagonal; A G and H D is the conducting rod, which is joined by G H for the sake of experiment; I is the conducting chain which connects the bottom of the conductor with the bottom of the jar, to complete the circuit with the discharging rod K. Now, if the jar represent a thunder cloud discharging its fluid in the direction K A towards the top of the house, it will be attracted by the rod, and carried by the metallic conductor A G H D to the earth, without injuring the building. But if the jar be charged again, and the square piece of wood be reversed, placing G H in the direction C B; when the jar is discharged, an explosion will ensue, and the piece of wood, which may be considered as a part of the building, will be driven out with considerable force, by the interruption which is given to the fluid in passing through the conductor.

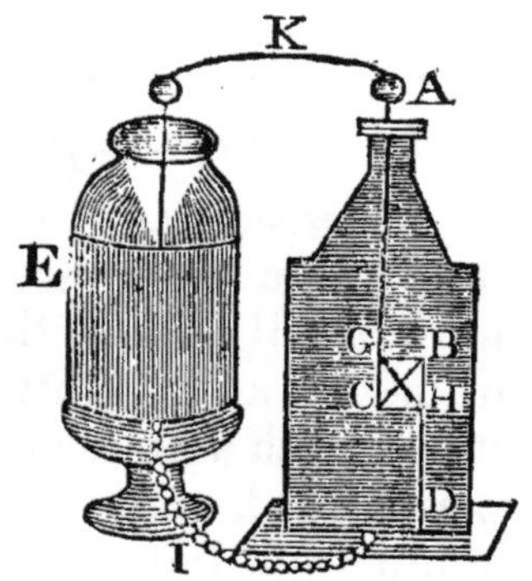

Remark. Dr. Franklin's electrical conductors, or lightning rods, are thus demonstrated to be a security against lightning, provided the lower end be set in the ground where they are always in moist soil, and provided the points are tipped with a metal which will not rust and become blunted.

A series of Leyden jars, presenting many superficial feet of metallic plate, may be made to produce powerful effects, by a sudden discharge of accumulated electrical power.

Illustration. In an *electric battery*, or a combination of jars, the accumulated fluid is capable of performing powerful experiments; but great care must be taken in using it, lest any person should chance to get into the electrical circuit, which would endanger his life if the battery were large. When the battery is used, it is likewise highly necessary to use the electrometer, to ascertain the height of the charge.

If a quire of paper be suspended by a string, and two ends of a conducting wire be brought near each side of it, and the circuit completed; on discharging the battery, the electric fluid will pierce a hole through the paper without putting it in motion.

Or, if a thick piece of glass be placed on an insulated stand, with a weight laid upon it, and the conducting wire of the machine be brought into contact with the ends of the glass; on discharging the battery, part of the glass will be reduced to powder, or if the glass be of tolerable thickness, it is sometimes coloured and shivered in a curious manner.

When the coated surface of the glass jars in the battery contains about thirty square feet, the fluid will melt brass wire of considerable thickness.

Galvanism.

Electric power may be excited, or have its equilibrium disturbed, by metallic plates alternating with dilute acids. When thus excited, it is called Galvanism.

Illustration. Make about a dozen copper cents, and as many silver quarter dollar pieces, as clean as possible. Join them in pairs, a cent and a quarter together, by pressing their surfaces together, and retaining them thus by waxing their edges all round the periphery. Pile these up in a column, with one of the metals always upwards, alternating with pieces of an old wool hat cut to the same form. This is called the Voltaic pile. If these pieces of wool hat be wet in a liquid, consisting of one part sulphuric acid, one part nitric acid, and sixty parts water, the electric power will be excited.

When the electric power is excited by the galvanic or voltaic process, the positive side is in the direction of the metallic plate which has the strongest affinity for oxygen.

Illustration. Let a trough be made of baked wood, and let it be divided transversely into half inch divisions, by double plates, each consisting of a plate of zinc and a plate of copper, with their edges soldered together. Care must be taken to have the same metal towards the same end of the trough in setting in all the plates; also to wax them in, so as to make each division water tight. Fill these interstices with dilute acid, as before described. The electrical equilibrium will immediately be disturbed, and will tend to restore itself in the direction of the zinc side of the plates; because it has a stronger affinity for oxygen than the copper. A circle may be formed by chains or wires, between the copper and zinc (or negative and positive) ends of the trough; and the most important experiments directed to be performed by the electric machine, may be now given.

Remark. The decomposing and deflagrating effects of the electric fluid when thus applied, belong strictly to the department of chemistry.

METEOROLOGY.

Remark. This department of natural philosophy, which is limited to the atmosphere and whatever floats in it, now occupies a much smaller proportion of works on philosophy than formerly. When conjecture on such subjects was about as well received as demonstration, meteorology was a fruitful field. But it has become extremely barren, since the philosopher feels under obligation to furnish proofs in support of his theories. No one subject furnishes more interesting phenomena, and no branch of natural philosophy is so little understood.

Aqueous Vapour.

Water is not exhaled into the atmosphere, in the state of vapour, by heat alone.

Illustration. Water is not converted into vapour by heat at a lower temperature than 212 degrees of Fahrenheit. The temperature of the air is rarely, if ever, above 100 in our latitudes.

Aqueous exhalation is effected chiefly by the attraction of adhesion, aided by heat.

Illustration. Expose a plate of ice, the sixteenth of an inch in thickness, to a dry atmosphere, when the temperature is below the freezing point ten or twenty degrees, and it will be wholly exhaled in eight or ten hours; but if the temperature were raised to 80 degrees of Fahrenheit, it would be exhaled in a less number of minutes.

The most dry, limpid, and pure atmospheric air always contains aqueous vapour.

Illustration. Pour sulphuric acid upon a little table salt in a wine-glass, which will disengage muriatic acid. The acid is itself invisible; but it will exhibit a dense smoky appearance, on account of its attracting together, in the form of a cloud, so many previously invisible atoms of water.

By some unknown process, invisible aqueous vapour becomes visible in the form of clouds.

Illustration. Watch the most limpid atmosphere, and clouds will be seen to form in the midst of the visible hemisphere.

Fine forms of clouds often precede each other in regular series.

Illustration. In fair weather during summer months, the *stratose* clouds, usually called fogs, often appear in the morning near the earth. After the sun shines upon them, they ascend in a state scarcely visible, and at length form the *cumulose* clouds. These are the bright shining clouds in brilliant heaps above, with apparently straight bases below, when viewed horizontally. They ascend still higher later in the day, and form the *cirrose* clouds. These have a fibrous flax-like appearance, and rise the highest of all clouds. At length they descend more or less, and become the *cirro-cumulose* clouds, by assuming a knotted or curdled appearance at first, and then becoming confluent. They either become stationary, producing rain, or break up, and their fragments become *cirro-stratose* clouds. These are the patches which have a stratified appearance when viewed horizontally; but they never approach the earth, like the first mentioned kind. No rain falls from this series of clouds, excepting while in the *cirro-cumulose* form.

These forms of clouds seem to be independent of all other forms, and of each other.

Illustration. The *nimbose* cloud, generally called the thunder cloud, as soon as it commences forming, begins to move pretty uniformly and steadily. At first it exhibits a heaped top, like the cumulose cloud; but as it advances in size, it shoots forth a kind of spray-like form from its uppermost heads. It usually produces rain, and breaks up soon after. The *villose* cloud is a kind of open fleecy cloud, which moves with great rapidity, often in a direction different from the clouds above. It is generally formed suddenly, and breaks up suddenly. The only remaining variety is the *cumulo-stratose* cloud. It is very rarely formed, and always appears to rise up in the horizon like the smoke from a furnace. Its top generally seems to pass into a *cirro-stratose* cloud above, and there spreads out like the top of a mushroom; it is therefore generally called the mushroom cloud.

All snow storms and settled rains proceed from the *cirro-cumulose*, and all hail storms and showers from the *nimbose*, clouds.

The vapour of clouds, by some unknown process, becomes condensed into liquid globules, and falls in drops of rain.

Illustration. As we know nothing of the process by which the limpid vapour of a clear atmosphere becomes a visible cloud, (though we can imitate it by muriatic acid, &c.); so we are equally ignorant of the process by which the visible vapour of clouds is formed into drops of rain. It is very manifest, that as soon as one minute drop is formed whose specific gravity is greater than that of the atmosphere, it commences falling. Whether the original drop formed in the cloud is enlarged or subdivided, while falling, seems not to be determined. A bucket of water dropped from the top of a steeple, would be subdivided into drops while falling through the atmosphere. But an extremely minute drop of water, as the first formed drops probably are, would commence falling very slow; because its specific gravity would hardly exceed that of the atmosphere. It would therefore probably condense and attract to itself other atoms of aqueous vapour, until its specific gravity became so great as to fall with that velocity with which we see it strike the earth.

By the rain-gage, (an instrument used for collecting and measuring the depth of rain falling in any place,) it appears that the quantity falling in a year is very variable in different places, and even in the same place. The average at Paris for several years was about 20 inches; London, 21; Liverpool, 37; Pisa in Italy, 43.

The vapour of clouds, and the first formed drops, may be in a state of congelation or frost.

Illustration. In the month of April in the year 1818, I was on the top of New-Lebanon mountain, on the east boundary of the state of New-York, in company with a full load of stage passengers, where I called their attention to the following phenomenon. The snow fell most rapidly on the top of the mountain, and for the distance of about the fourth of a mile along its descent. At that place, and throughout all the valley below, it rained. We stopped, and found ourselves able to perceive the precise limit of the termination of the snow, and of the commencement of the rain. Nothing was left for conjecture; for we saw the very flakes of snow melt into rain at that precise limit. On descending the hill a little further, where it rained copiously, and looking upward, the termination of the snow appeared like a vast snow-bank hanging over our heads. But after descending into the valley of New-Lebanon, the snow over our heads appeared like a common cloud in an ordinary steady rain, though no change had probably taken place in it.

It is well known, that the temperature of the atmosphere in the higher regions, where clouds often float, is generally below freezing, when at the same time it is many degrees above freezing in the lower regions. In the case just described, it is manifest that the flakes of snow were melted into rain by the warm atmosphere a little lower than the top of the mountain, while on the top it remained frozen in a temperature below freezing point.

The equilibrium of the electric fluid is often disturbed in clouds, by some unknown cause; and violent pulsations in the atmosphere are produced by its restoration.

Illustration. It has been already shown that bodies may be positively and negatively electrified by disturbing the equilibrium, and the effects of its restoration have been shewn artificially. In the same manner, the electric fluid probably passes from a cloud positively electrified to a negative one, and sometimes to the earth. Violent percussions are made upon the atmosphere by its motion, producing sounds called thunder.

Wind.

Winds are chiefly caused by variation of temperature; and they commence where the air is rarefied by heat, and blow in the direction of the point where they commence.

Illustration. Let a fire be kindled in a heap of dry straw, or other highly combustible substance, when the air is perfectly calm. The wind will soon commence blowing towards the fire, and the air mostly ascends as it becomes rarefied: part will pass on to a greater or less distance, however. The next portion of air will follow the last while seeking its equilibrium pressure, until the wind is extended back to a great distance in the direction from whence it proceeded. On this principle, winds blow towards side hills and other places, which, by their position, peculiar constituents, &c. become highly heated. The trade winds, monsoons, &c. may be explained on this principle.

In the northern latitudes, when northwesterly winds commence, the air soon becomes dry and cold, on account of its moving obliquely downwards from the higher regions.

Illustration. The sudden change of temperature which follows a change of wind from any other point of compass to the northwesterly, cannot be produced by the progress of the wind from the cold regions of the north; for the change is effected before the wind has time to travel far enough to produce any effect. But in breaking down obliquely from the cold regions a few miles above the earth, the temperature is soon changed. As the air is charged with vapour near the earth, almost to saturation, after it has remained a while in that situation, very little vapour is taken up by it. But when air breaks down from the upper regions, where it necessarily contains less vapour even in proportion to its density, it takes up vapour with great avidity, until it has its equilibrium proportion.

Winds sometimes take a spiral ascending motion, when they are called whirlwinds.

Illustration. On the 2d day of October, 1820, I was in the town of Sharon, in Schoharie county, N. Y. in company with the Hon. Farrand Stranahan and Mr. Pohlman, principal of Hartwick academy, where we saw, in the midst of a very calm atmosphere, at 2 o'clock P. M. the effects of a remarkable whirlwind. It carried up a very dense cloud of dust, leaves, sticks, &c. in a perfectly straight, even column about six feet in diameter, to a great height. By comparing the height of the trees with that of the column, we judged its height to be more than three hundred feet. Limbs of trees three inches in diameter, and from one to three feet in length, together with stones an inch in diameter, were carried up almost to the top. Water-spouts are probably carried up by similar whirlwinds.

Remarks. The velocity of winds is very variable —from an almost imperceptible motion, to a velocity of one hundred miles an hour.

The following table from Cavallo was made by Mr. Rouse, and has been sanctioned by high authority.

Hardly perceptible	1 mile in an hour.
Distinctly perceptible	2 to 3 miles an hour
Pleasant breeze	4 to 5
Pleasant gale	10 to 15
Brisk gale	20 to 25
High wind	30 to 35
Rapid wind	40 to 45
Tempest	50
Stormy tempest	60
Hurricane	80
Rapid hurricane	100

The strength of winds is measured by various contrivances, called wind-gages, or anemometers; such as a swinging pendulum, which is so contrived as to mark the fartherest point to which it vibrates. It has been proved, that one foot area has presented a force equal to about fifty weight of resistance, in a violent hurricane.

Winds may be reduced into three classes, called variable, periodical, and general.

Variable winds are those which are not subject to any particular period, either in duration or return.

The stated or periodical winds are such as return at certain times; these may be divided into two kinds, viz. the sea and land breezes, which are produced by the diurnal motion of the sun, and blow alternately from the sea and land; and monsoons, which are caused by the annual revolution of the sun, blowing one way for a certain number of months, and the opposite way for the rest of the year.

General winds, which are usually called trade winds, blow always nearly in the same direction, as those between the tropics in the Atlantic and Pacific oceans.

The annual and diurnal revolution of the sun may be the general cause of winds; but this hypothesis can by no means sufficiently explain the phenomena, without having recourse to some other aid; as these causes could only produce regular winds, the progress of which would correspond, and be connected with the seasons.

As the sun appears to be continually shifting to the westward during a diurnal revolution, the lower air becomes attenuated by his rays, and the tendency of the whole body is towards this rarefied passage, which produces a northeasterly and southeasterly wind to about 30 degrees on each side of the equator, called trade winds. As this motion is communicated to a vast ocean of air, the current continues during the night, till the sun appears again to give fresh impulse, and restore the motion that was lost in its absence, which causes the wind to be perpetually in this direction. The rarefied air which ascends near the equator, returns in a northeast and southeast direction in the upper regions; just as the warmed air of a room rushes out at the top of a door, so as to blow a candle blaze outwards, while it is blown inwards at the bottom.

If the surface of the globe were covered with water alone, the winds would be perpetually the same as the trade winds in the Atlantic and Pacific oceans; but the large continents of land receiving a greater degree of heat than the water during the day, they communicate it to the air above them, which becoming more rarefied than that part over the sea, the denser air passes towards the land. This accounts for those westerly, instead of easterly winds that blow towards the coast of Guinea from some distance on the sea.

The sea and land breezes in the West Indies no doubt arise from the same cause. The breeze from the sea to the land begins to appear about nine o'clock in the morning, and keeps gradually increasing till noon, and dies away about four or five in the afternoon; about six in the evening it changes into a land breeze, which continues blowing towards the sea till near eight the next morning.

These changes may be accounted for according to the preceding principles; for as the heat takes more effect on the land than on the water during the day, the air over the land becomes more rarefied, which causes the cooler air to rush in to keep up the equilibrium. In the evening, as soon as the sun is set, the dews come on so excessively, that the air becomes suddenly cooled, consequently more dense than that which is over the water, and this causes the air to press from the land to the water, which produces the opposite current.

The cause of the monsoons, or periodical winds, is owing to the course of the sun northwards of the equator one half of the year, and southwards the other. While it passes through the six northern signs of the ecliptic, the various countries of Arabia, Persia, India, and China, are heated, and reflect great quantities of the solar rays into the atmosphere; by which means it becomes greatly rarefied, and the equilibrium is destroyed: in restoring it again, the air not only rushes in from the equatorial parts southwards where it is colder, but likewise that from the northern climes must necessarily have a tendency towards those parts, which produces the monsoons for the first six months. Then, during the other six months, whilst the sun is traversing the ocean and countries in the southern tropic, it heats and rarefies the air in those parts, consequently causes the equatorial air to alter its course, to veer quite about, and blow from the opposite points of the horizon.

These are the general affections of constant and regular winds; but the whole are subject to variations and exceptions, on account of different circumstances, local or otherwise.

Luminous Meteors.

The ignis fatuus is probably phosphuretted hydrogen combined with a gelatinous substance.

Illustration. Every one who has attended chemical experiments, knows that phosphuretted hydrogen gas burns spontaneously with a kind of puffing explosion, like the ignis fatuus, (called also Jack o' lantern, and Will o' the wisp.) It is also produced about burying grounds, and places where the dead cattle and horses of a village are thrown. As the bones of animals contain phosphorus, nature probably has a method of disengaging it from the bones, and then decomposing water with it by the aid of the lime in the bones, on the same principle by which the chemist performs the same operation in the laboratory. This explanation is far from being satisfactory; but it is the best which the present state of the science has furnished.

Other luminous meteors, such as are called shooting stars, &c. may be explained as well upon the foregoing principle, as upon any other hitherto suggested. All is but conjecture at best, however.

The aurora borealis is probably caused by electrical excitement in the upper regions, where the atmosphere is extremely rare.

Illustration. Exhaust the atmospheric air, by heat or with the air-pump, from a glass matrass, or other glass vessel resembling a florence flask, and then stop it tight by a cork or good stop-cock. Rub the vessel in the dark with an amalgam on leather, as directed under electricity; and it will appear luminous within, being filled with flashes of light greatly resembling the aurora borealis.

LIGHT.

Remarks. Light is that power by which objects are made perceptible to our sense of seeing, or the sensation occasioned in the mind by the view of luminous objects.

Light, like many other effects in nature, the principle and essence of which exceed the bounds of human understanding, has, for many ages, been a subject of much speculation and hypothesis. It has been considered as a mere quality of particular bodies, or a fluid medium by which the vibrations of luminous bodies are carried to the eye; but Newton demonstrates it to be an absolute body, composed of infinitely small particles of matter, which issue by a repulsive force from luminous bodies with wonderful velocity, diverging in right lines in all directions.

Motion of Light.

Particles of light are so extremely minute, as to pervade some of the hardest, and very compact, bodies.

Illustration. That the particles of light are infinitely small, may be reasonably inferred from their penetrating the densest bodies. The pores of glass, crystal, or a diamond, cannot stop the subtilty of light; and yet the greatest collection of this matter has never been found to have any sensible weight, even in the finest scale. Considering the immense velocity of light, if the particles were not infinitely small, and probably placed at a considerable distance from one another, the pressure on the eye would be insufferable.

Rays of light move in straight lines.

Illustration. It appears evident that the rays of light are emitted in right lines, from the shadow which is thrown behind those bodies on which they fall, as the corresponding parts of the substance and shadow form right lines with the source of the ray.

Light is not instantaneous, but progresses at the rate of about two hundred thousand miles in a second.

Illustration. The velocity of light has been considered instantaneous; nor can its passage from one visible object to another be marked by any difference of time, although the following discovery of Roemer's, supported by Cassini and others, sufficiently proves a progressive motion. He observed that the eclipses of Jupiter's satellites varied 16½ minutes in time in some particular situations of the earth in its orbit, being 8¼ minutes sooner than the calculated time when the earth was nearest to the planet, and 8¼ minutes later than the tables when the earth was in the opposite part of its orbit. From this observation, Cassini and others have concluded that the difference of time proceeds from the progressive motion of light in passing the orbit of the earth. This subject may be further explained by the following diagram.

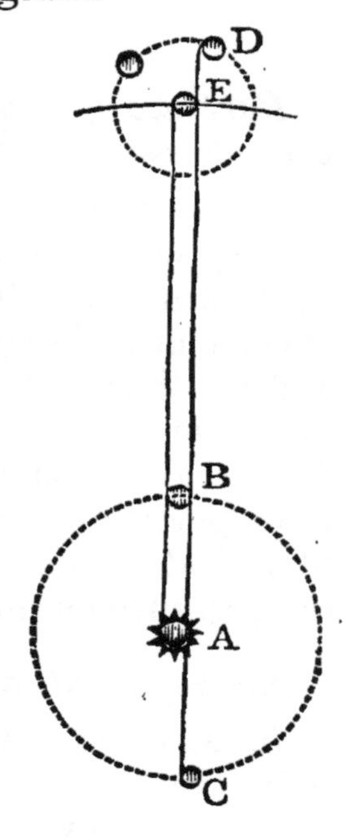

Let A be the sun, or the centre of the system, from which the tables are calculated; B C the earth's orbit; E Jupiter, and D one of his satellites just entering his disk. Then an observer at A would find the time of immersion coincide with the tables; but at B it takes place 8¼ minutes sooner than at A, and at C 8¼ minutes later; which, taken together, gives a difference of 16½ minutes for the progressive time of the reflected light of the satellite in passing from B to C, the diameter of the earth's orbit. Now, if we take the diameter of the earth's orbit at 190 millions of miles, and divide it by 990, the seconds in 16¼ minutes, the result will be about 200,000 miles, which is the wonderful velocity of light in a second of time.

As the sun is placed in the centre of the earth's orbit, the distance of these two bodies from each other is equal to the semi-diameter of the earth's orbit, or 95 millions of miles; so that the particles of light are transmitted from the sun to the earth in 8¼ minutes, which is near two millions of times swifter than the velocity of a cannon ball, supposing it to fly at the rate of 450 miles in an hour.

Colours.

Light consists of seven kinds, which are distinguished by their degrees of refrangibility, and by the different sensations of colour which they excite.

Illustration. The sun's light consists of rays, which have a considerable inequality of refraction as they are transmitted through the medium of the atmosphere, and impress a sensation of colours as they are more or less bent in their course, which we call violet, indigo, blue, green, yellow, orange, and red. Those rays which pass the most directly, or are the least refrangible, produce red; the next rays in refrangibility produce orange, then yellow, &c. The rays which produce violet are the most broken, or have the greatest refrangibility.

These are all the primary colours in nature; but by blending or mixing different rays together, they mutually alloy each other, and constitute intermediate colours, and shades of colours of every description.

Whiteness is a peculiar production: it is not composed of any particular ray of light, but produced by a copious reflection and due proportion of the rays of all sorts of colours.

Blackness proceeds from the peculiar quality of a body which stifles and absorbs the rays of light that fall upon it; so that instead of reflecting them outwards, they are reflected and refracted inwards, till the incident rays are lost.

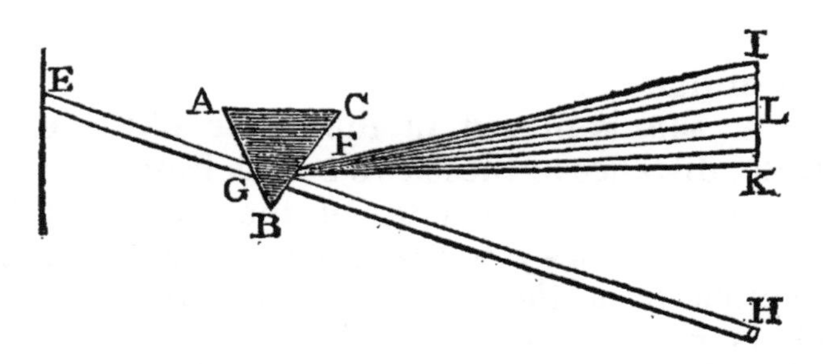

Let A B C represent a glass prism, and E a small hole in the window-shutter of a darkened room, by which the pencil of rays E G enters and falls on the side of the prism at G. If the medium of the glass did not obstruct the rays, they would pass on in a straight line and parallel direction to H, and there illuminate a small circle in the side of the room H I; but when the rays enter and pass out of the denser medium of the glass, they are refracted or bent towards L; therefore, if the rays were equally refracted by the prism, they would pass on from F to L in parallel lines, and enlighten a circular spot at L, similar to that which was formed at H. But if the pencil of light be composed of rays which are not equally refrangible, then those which are the least refrangible will fall nearest to the right-lined direction E G H, and those that are the most refrangible will be the most distant, and the intermediate degrees of refrangibility will issue in different rays between the two extremes. This accords with experiment; for after the rays quit the side of the prism F, they diverge according to their refrangibility, and form an oblong spectrum, variously coloured, on the side of the wall between I and K; the lower part of which, being the least bent, produces a lively red colour; this changes by gradation into an orange, thence into a yellow, and as the rays rise higher, into a green, blue, indigo, and violet, which is the most distant, as being the most broken.

Remark. There is a remarkable analogy between colour and sound; for it is found that the divisions of the uncompounded colours on the spectrum agree with the different divisions of a musical chord.

As an additional proof that the refrangibility of the sun's rays is various, and that the different rays reflect their own colours; if the spectrum be received on a perforated board, so that the uncompounded colours may pass distinctly through the hole, it will be found that they still preserve their individual colour, whether it be received on a sheet of white paper behind the hole, or be again refracted by another prism upon some other surface.

Those rays which are the most refrangible, are also the most reflexible.

Illustration. If the prism A B E be placed in a darkened room, in such a manner that the pencil of light which passes through the hole D may be reflected from the point E in the base; the violet rays will be first seen reflected in the upper line E F, and the other rays will continue their refraction through E C and E G, &c.; but if the prism be gently moved on its axis, the indigo will be reflected after the violet, then the blue, green, and so on till the red be reflected, which is the last.

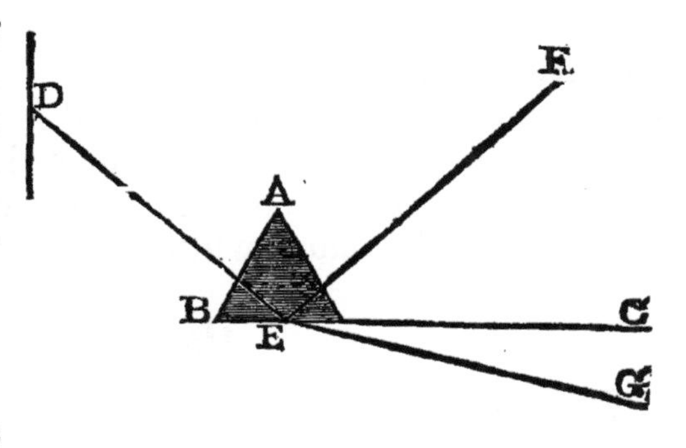

Colouring is not inherent in coloured bodies, but depends on the arrangement of the constituent atoms of bodies.

Illustration. As the rays of light differ both in refrangibility and reflexibility, in accounting for the different colours of bodies, it is manifest that different bodies are endued with a power or aptitude to reflect the rays of one particular colour, and to imbibe the rest. This opinion is supported by experiment; for if any body be placed in the uncompounded light of the spectrum, so that a pure and distinct colour may fall upon it, the body will appear of the same colour, only with this difference, that it will reflect that colour most brightly which is the same as the body itself reflects in full light. But as all bodies reflect other colours in some degree, besides the predominant colour which belongs to them, they cannot appear so full and clear as the colour is seen in the spectrum, where it is uncompounded ; but they reflect their colour feebly and weak, in proportion to the compound of the rays which are reflected.

The most conclusive experiments belong to the department of chemistry. For example, take two colourless liquids, as a solution of pearlash and corrosive sublimate : though limpid before, they will be orange-coloured after, being mixed. A limpid solution of pearlash mixed with a limpid solution of sugar of lead, will be snow white.

Optics.

Rays of light, when passing out of a rare into a dense medium obliquely, tend to enter perpendicularly ; when out of a dense into a rare medium, tend to pass out still more obliquely.

Illustration. Rays or pencils of light pass in right lines from luminous objects, each carrying an impression to the eye of that part of the object whence it was emitted. When the direction of the ray keeps in the same medium, the object, image, and eye, will be in a straight line ; but when the ray is transmitted obliquely through different mediums, such as air, water, glass, or any other transparent body, as it enters into each, it changes its direction, inclining more or less to the perpendicular of the medium, according to the density of the body. This deflection (usually called *refraction)* is supposed to proceed from the attraction of the denser medium, which acts in right lines perpendicular to the surface : thus the attractive power impedes the rays in their oblique course, and draws them towards the axis of the medium. When rays fall perpendicular, they have no refraction.

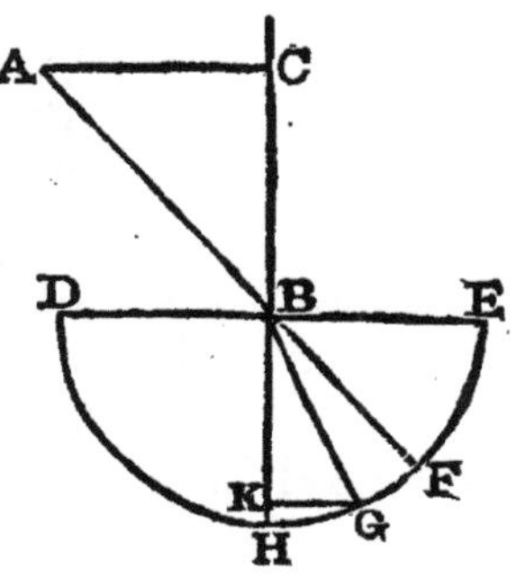

D H E represents a vessel filled with water, or any transparent body denser than air, and D B E is its surface. The line A B is the course of the ray of light from the object or radiant at A, which is called the line of incidence, and B is the point of incidence; the line G B is the line of refraction, or the course of the ray, when the direction is changed at B, by entering into a denser medium than that of the air through which it has passed; the right line D E, or the surface of the medium, where the lines of incidence and refraction meet each other, is called the plane of refraction; B H, the perpendicular to that plane, is the axis of refraction; and the continuation of the same line, C B, above the surface of the medium, is called the axis of incidence. The angle A B C, which is formed by the line of incidence and the perpendicular C B, is called the angle of incidence; and A C is the sine of the angle of incidence. The angle H B G, which is formed by the line of refraction and its axis, is the angle of refraction; K G is its sine; and the angle G B F, which is the difference between the angles of incidence and rèfraction, is called the angle of deviation.

It may be practically found, by the following experiment, that light, in passing out of a rare into a denser medium, approaches nearer to the perpendicular, and that it recedes in passing from a denser to a rarer.

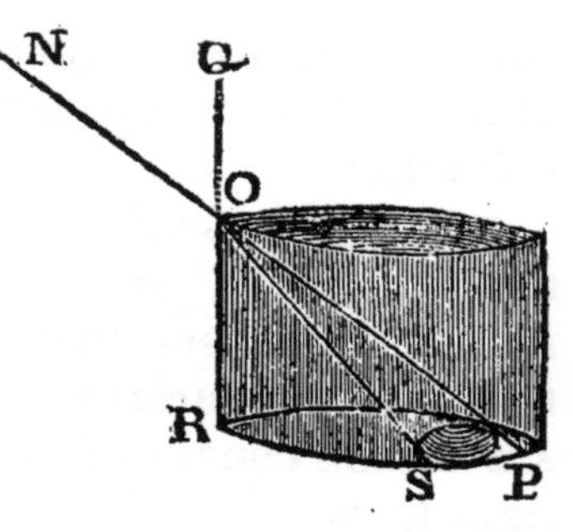

A piece of money being placed at the bottom of the cylindrical vessel O R, let the eye of the observer be so placed at N as just to lose sight of the object at P, or so that a ray shall pass from the remote part of the piece to the eye in the direction P O N. Whilst the eye continues in this situation, if the vessel be filled with water, the object will become visible; for the rays which pass from S in a right line will be bent at O, the surface of the water, and fall into the eye at N, or be refracted from the perpendicular O Q. It may here be necessary to observe, that impressions are received in the eye by the rays which proceed from the object; so that in the above experiment, when the vessel is filled with water, the rays pass from a denser into a rarer medium, consequently recede from the axis of refraction when they enter the eye. If we consider the ray as issuing from N, that is, from the rare into the denser medium, it will pass in the direction O S, approaching the perpendicular; and as the water is drawn out of the vessel, the ray will recede till it falls into the straight line N O P, at the outer edge of the coin.

Remarks. The course of rays, or pencils of light, is divided into three kinds, viz. parallel, converging, and diverging rays.

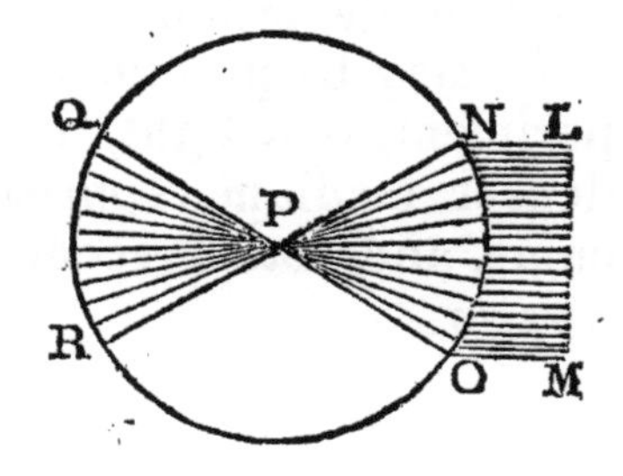

Rays are called *parallel*, when they would pass to infinity at equal distances from each other, as N L O M. When rays of light issue from bodies at immense distances, they are considered as parallel to one another; so that those rays which proceed from the sun and pass through our atmosphere, are taken as parallel, though each point of the sun is a radiant, diverging rays of light on the earth; but as the distance is so immense, and the angle of divergency so infinitely small, the rays may be fairly considered as passing in the same parallelism.

Converging rays are those which, in passing from a rare to a denser medium, are refracted or bent towards the perpendicular, and meet in a common centre, called the *focus*, as N P O.

Diverging rays recede from the axis or perpendicular, in passing from a dense into a rarer medium; or when converging rays have crossed each other in their focus, or burning point, then they diverge or pass off from one another, as Q P R represents in the figure.

Refracted rays of light have the sines of the angles of incidence, in the same medium, in regular proportion to each other.

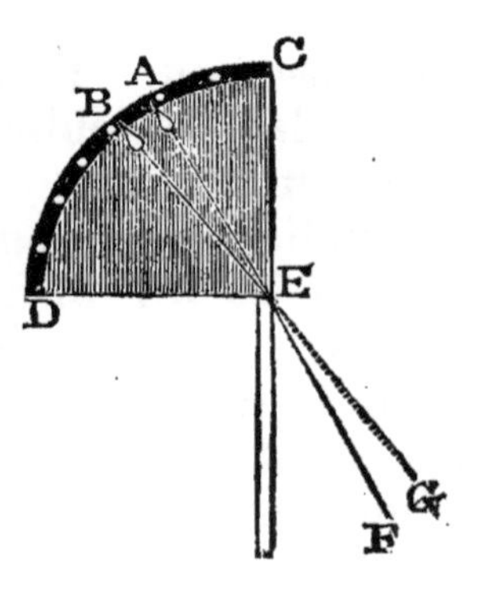

Illustration. Let the quadrant C D E, which is graduated on its circular edge, have two moving indices, A and B, that turn in the point E, and let A E be prolonged to F; then set the index A to 15 degrees on the scale, and B to 19¼°, and bring the edge of the quadrant D E to the surface of a vessel of clear water, immersing the lengthened part of the index A F: now the immersed part F E will appear bent to G, which is in a right line with the index B E. In like manner, if A be removed to 30°, and B to 41½°, the refraction of E F will bring the apparent place of the limb in a right line with B E. Thus the angle of refraction may be found to any line of incidence, either out of a rare into a denser medium, or from a denser into a rarer; by placing the longer index to the angle of incidence on the quadrant, then immersing it in water, and afterwards moving the shorter index till it apparently coincide with the refracted limb of the other, and the number of degrees on the scale opposite to the shorter index, will be the angle of refraction, whether the incidental ray issue from a denser or rarer medium.

But as the sines of the angles of incidence and refraction, between the same media, are in a constant ratio, if the angles of incidence and refraction of one incidental line be given, any other may be known by finding the sines, and saying,

As the sine of the known angle of incidence
Is to the sine of its refraction,
So is the sine of any other angle of incidence
To the sine of its refraction.

When rays of light strike upon the polished surface of a hard body, they are turned back in such a direction that the angles of incidence and of reflection are equal.

Illustration. Rays of light may either fall perpendicularly or obliquely upon a mirror or speculum.

When the ray falls perpendicularly upon the surface, as D F on A B, the ray is reflected back again in the same direction; but when it falls obliquely, like C F, on the plane, it is reflected from the surface, and passes in the direction F E, making the angle of reflection E F D exactly equal to the angle of incidence C F D. Consequently,

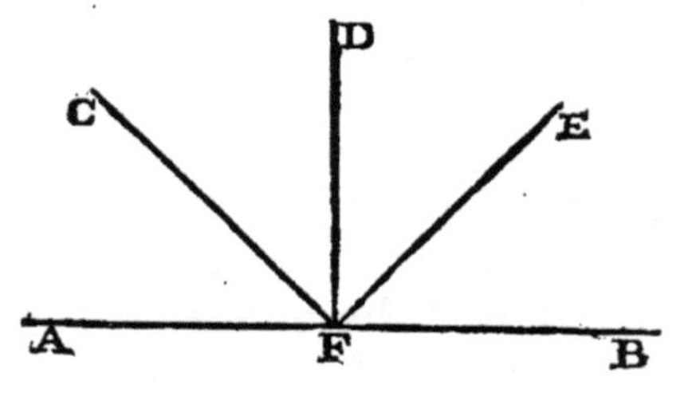

Parallel rays falling on a plane mirror still keep parallel after reflection; for as the angle of incidence is equal to the angle of reflection, and as the incidental rays E F and H I, are parallel to each other, the reflected rays G I and C F, will likewise be parallel.

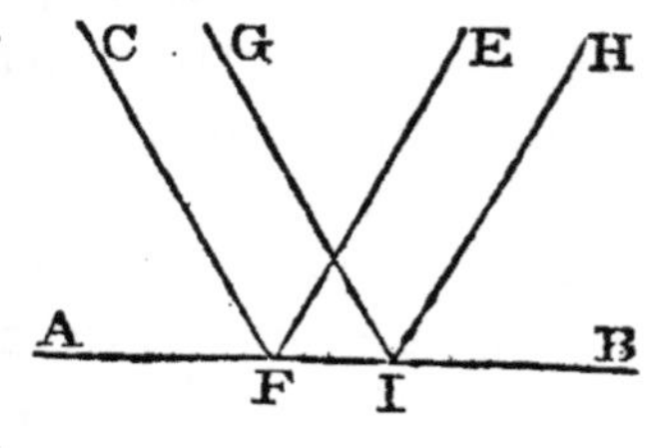

Diverging rays have the same divergency when

they are reflected, as they would have at an equal distance if continued in right lines.

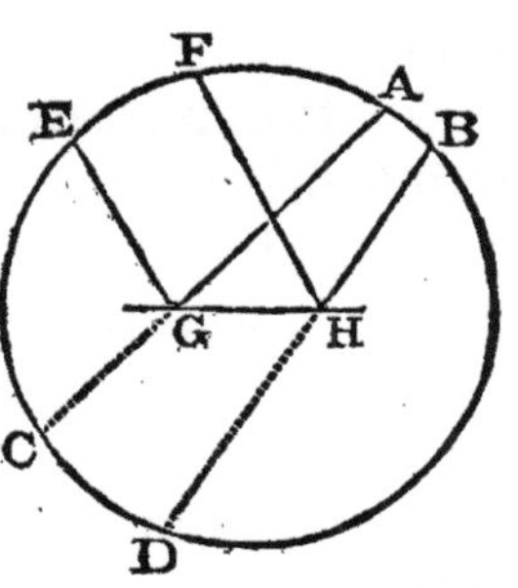

That is, if two diverging rays A G and B H be reflected to E and F, the arc E F will be equal to the arc D C, which would have been the position of the rays, supposing them to have passed in a right lined direction.

Converging rays which are reflected, have the same convergency at an equal distance, as if they passed in straight lines.

Thus, as the converging rays I N and K O have their angles of incidence equal to their angles of reflection, they converge to the point M, at a distance equal to L, where the rays would have met in an uninterrupted course.

By *spherical* or *convex surfaces,* parallel rays are rendered divergent. Every spherical surface may be considered as composed of an infinite number of straight lines.

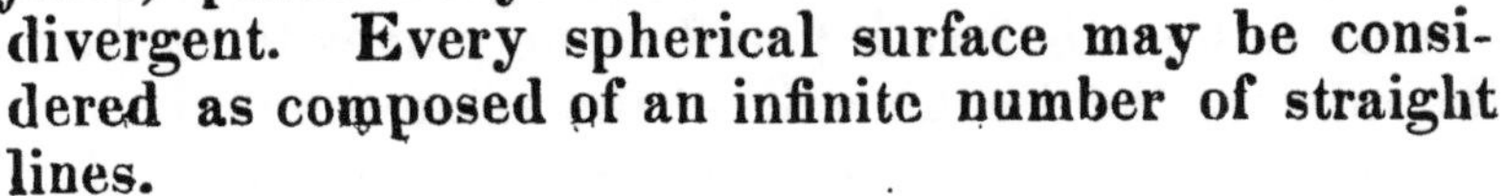

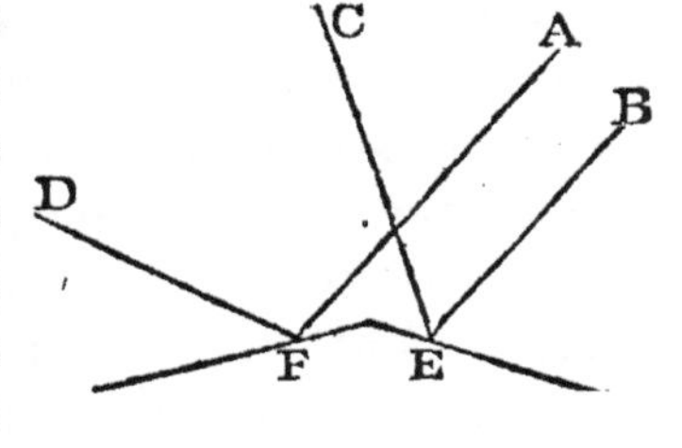

Suppose E F two of those straight lines which form part of the convex surface of a sphere, and that the rays A F, B E, fall parallel on the points E and F, then it is evident, that as these lines are reflected in equal angles from the oblique lines E and F, they will diverge, and lose their parallel direction.

By considering converging rays in the same manner, it appears that they will become less convergent, and diverging rays more divergent.

In concave surfaces, parallel rays are made convergent.

For as this surface, like that of the convex mirror, may be supposed to be formed by an infinite number of right lines, the parallel rays K H and I G, which fall on the oblique lines G and H, are reflected in equal angles through the lines G L and H M, tending towards each other till they meet in the vertex.

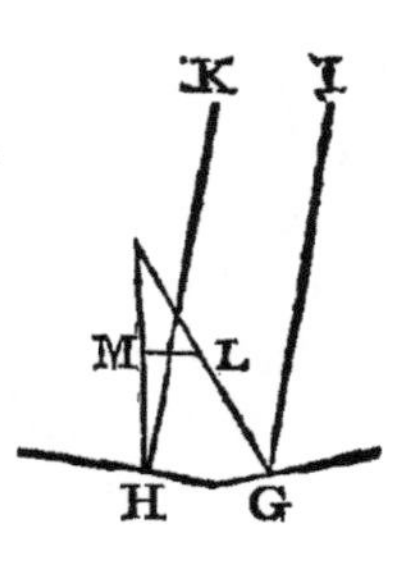

In like manner it may be shown, on the same superficies, that rays already convergent become still more so, and that diverging rays are made less divergent.

From the foregoing principles, it will be easy to comprehend the effects of mirrors, and account for the principal phenomena which may occur, either with those that are plane, convex, or concave.

A *plane mirror* does not alter the figure, or change the size of objects; but the whole image is equal and similar to the whole object, and has the same situation on one side of the plane that the object has on the other.

A spectator will see his own image as far beyond a mirror as he is before it; and as he moves to or from the mirror, the image will at the same time advance to, or recede from him, on the opposite side, but seemingly with double velocity, because the two motions are equal and contrary.

If a person view himself in a plane looking-glass, he will see himself completely, at any distance, in a part of the glass the length and breadth of which are equal to half the length and breadth of the corresponding parts of his body; for as the image appears as far behind the glass as he stands before it, the part of the mirror on which the rays fall will be equal to half the length or breadth of the object, or the rays will only spread half as much as they would do at double the distance.

In a *convex mirror*, the image always appears smaller than the object, and the diminution increases as the object recedes. This will be easily understood, when it is considered that the reflecting convex surface of a mirror renders incident converging rays less convergent.

The image does not appear so far behind a reflecting convex mirror as in a plane one; for the diverging rays are reflected more divergent, consequently they have their imaginary focus much nearer, which makes the image appear nearer to the surface of reflection.

A *concave mirror* differs from the two preceding. It only shows bodies erect when the object is placed between its real focus and the mirror; then the rays are rendered convergent, and the image appears larger than the object; but when it is placed beyond the focus of the mirror, the rays cross each other at the focus, and the image appears inverted.

Application of the general law of refraction of rays of light to lenses.

A lens is a piece of glass or crystal so formed, that rays of light, in passing through it, have their direction changed. It either converges them to a point or focus beyond the lens, or diverges them as if the rays had proceeded from a point before it, or brings converging and diverging rays parallel to each other.

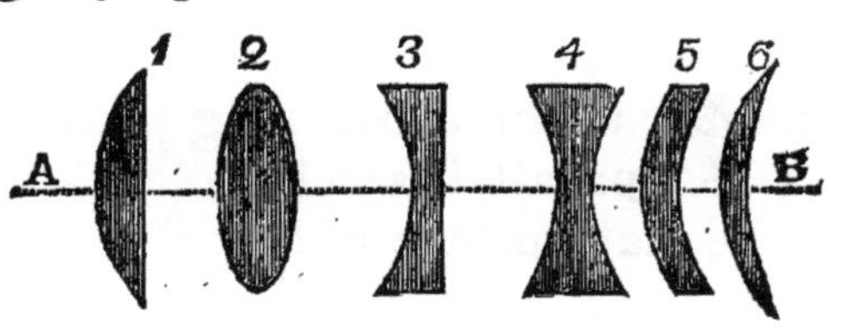

The lens marked 1, is called *plano-convex*, having one of its sides plane, and the other spherical, which forms the segment of a sphere. 2. Is *double convex*, having both sides the same, and is like two equal segments of a sphere joined together. 3. Is a *plano-concave* lens; one of its sides being flat, and the other hollow, such as would be represented by the impression of a small part of a sphere in soft wax. 4. Is *double concave*, having both its sides equally hollow. The fifth is called a *concavo-convex*, having one of its sides concave, and the other convex. The 6th is called a *meniscus,* having one side convex and the other concave, like the 5th; but the convexity exceeding the concavity, the two sides meet so as to form an edge.

The line B D, which passes through the middle of the lens perpendicular to the sides, is called the axis; the two points where it enters and passes out of the lens, the vertices; and the distance between them, the diameter. The focus, either of converging or diverging rays, is situated somewhere in the axis of the lens.

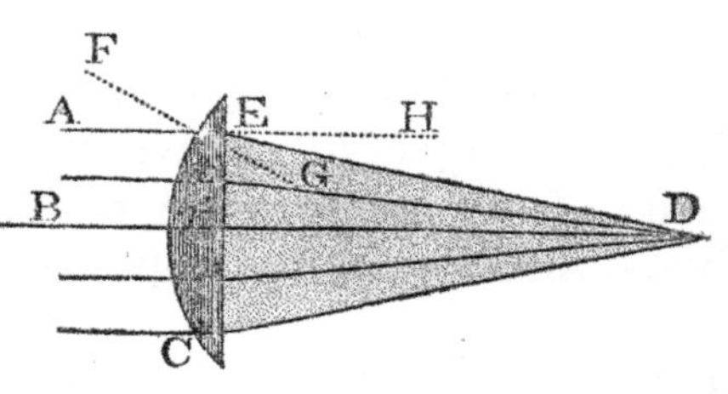

If the ray of light A E fall upon the plano-convex lens at E, it will not pass on in the right line A H; but as it is transmitted through the glass, its course will be refracted into the line E D, approaching towards the perpendicular of the convex side F G. Another ray falling on C parallel to A E, and equidistant from the axis B, will be refracted in like manner, and will converge and meet A E in D, the focal point of the lens. All the intermediate rays which fall between E and C converge in the same way, only the rays will be less refracted as they approach towards B, till they fall into the centre, and pass in a right line through the diameter to the converging point D.

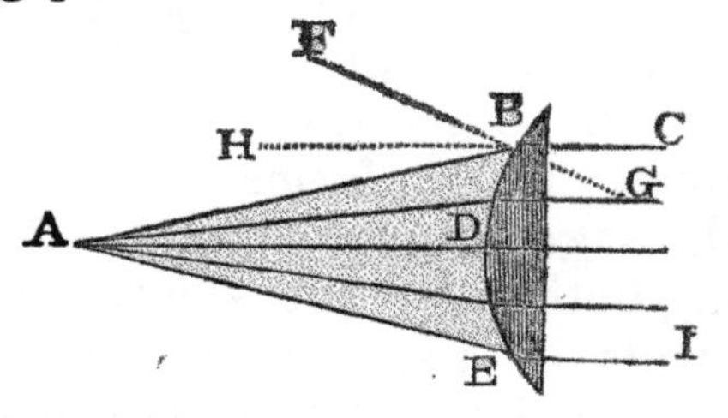

When parallel rays fall on the flat surface of the same figure, they will tend from the perpendicular F G of the convex side, and converge with the pencil of rays from I E and the intermediate parallel rays to the point A.

Considering the two mediums, air and glass, through which the rays pass, it appears that a ray from the spherical surface of the lens, whether in passing through the denser medium of the glass, or the rarer medium of the air, still converges to a point somewhere in the axis ; therefore, if both sides of the lens be spherical, the convergency will still be greater, and the rays will meet in some point nearer to the centre of the lens.

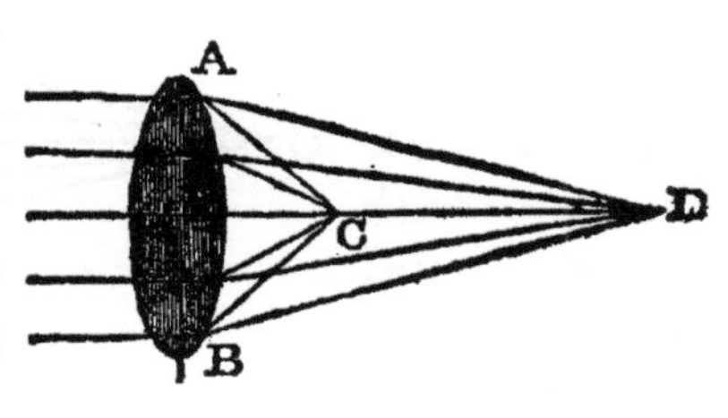

Consequently, if A B be a double convex lens, the parallel rays which fall upon it will converge to the point C, by the double convexity of its sides ; whereas if one of its sides had been flat, as in the preceding example, the converging point would have been extended to D.

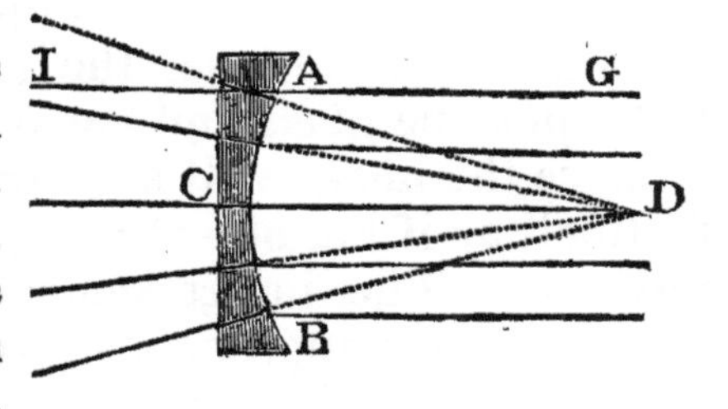

If A B be a plano-concave lens, and C D the axis ; let the ray A G fall upon the lens at A, then it will be refracted in passing through the glass, and diverge from the direct line G I into A D, approaching towards the perpendicular of the concave side. The ray at B being equidistant from C, will diverge in the same manner, as well as the rest of the intermediate rays, in proportion to their distance from the axis of the lens.

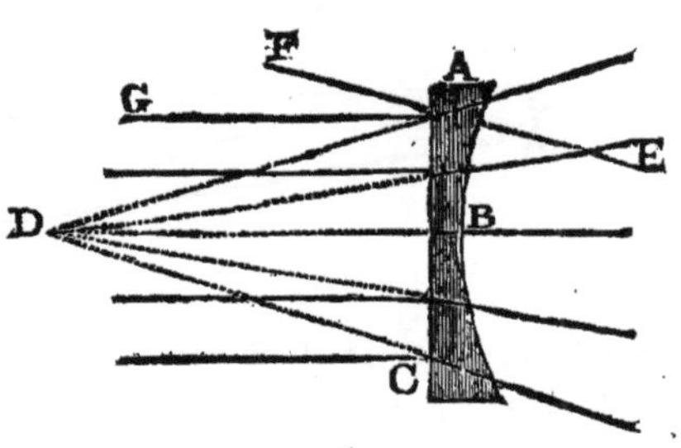

Let the rays fall on the flat side of the lens : after having passed through the denser medium of the glass, they will diverge on the opposite side as they pass into the rarer medium of the air, and the ray G A will be refracted from the perpendicular F E, as if it proceeded from D ; likewise C, and all the intermediate rays, will diverge in proportion to their distance from the axis.

Consequently, if the rays equally diverge from the

hollow surface, either in passing from or into the lens, there will be an equal divergency from both sides of the double concave lens A B; and those rays, which would have diverged in a plano-concave lens as coming from C, now diverge with the additional concave side, as if they issued from the virtual focus E, in the axis of the lens C D.

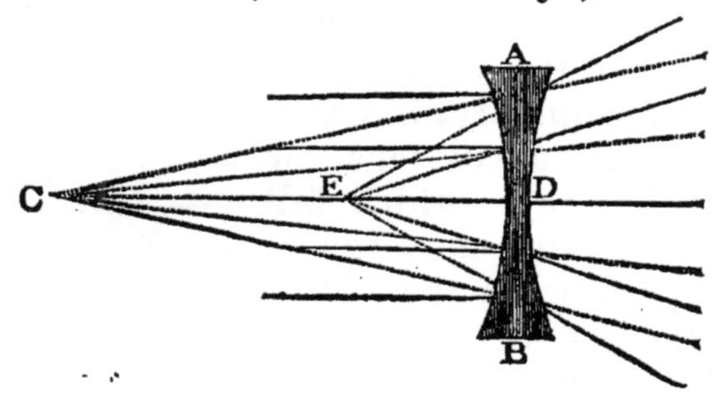

When the radiant, or object, is at a considerable distance from the lens, the rays issuing from it will fall upon the glass and converge to the focal point; whence the image will appear inverted, but clear and distinct, as if the object was placed there in the same position. The image removes further from the lens as the radiant approaches; so that when it is brought within the focal point, the rays diverge to infinite distances, passing through the lens in parallel lines.

If A B represent a double convex lens, and C D an object at a considerable distance from the glass, the

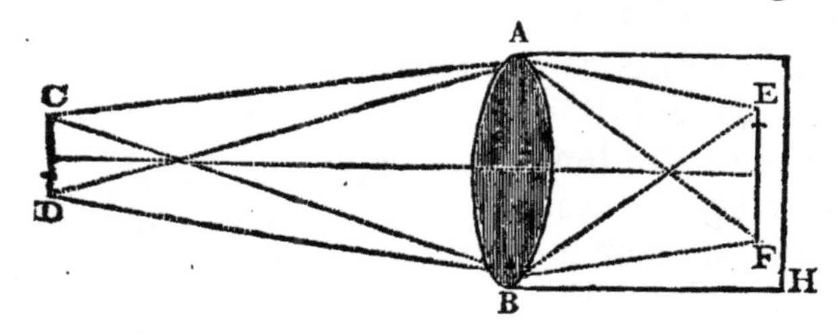

rays D A and D B issuing from the point D, will be refracted at A and B, and converge to E, and form one extremity of the object; in the same manner C A and C B, in passing from C, will converge to F, the other extremity; then if we conceive an infinite number of rays passing from every other point of the object C D, they will cross each other and meet somewhere between E and F, the foci of the lens, forming a complete image of the object reversed, and the linear magnitude of the object and image will be relatively as their distance from the lens.

Application of the general law of refraction to the camera obscura, magic lantern, burning glass, and telescope.

Camera obscura. This machine is formed on the above principle; for if G H represent a darkened room, and A B a lens fixed in the side of it, the object C D, with all its shadows and colours, will be distinctly seen on a white surface placed in the focus of the glass, but in an inverted position. As the appearance of inverted objects is unpleasant to the eye, if the lens of the camera be placed in a short tube on the top of a small building, and the image of the objects be reflected through the lens by an inverted mirror placed above it, the picture will be presented in a proper position upon the receiving table in the focal point of the lens, giving the most beautiful and animated representation of all the surrounding objects in their own colours.

Magic lantern. This amusing machine is made to magnify small pictures, which are painted upon glass, and to throw the shadow upon the side of a darkened room. It is principally formed by a convex lens.

A lighted candle or lamp is placed in the inside of a square box, which has the tube G B projecting from

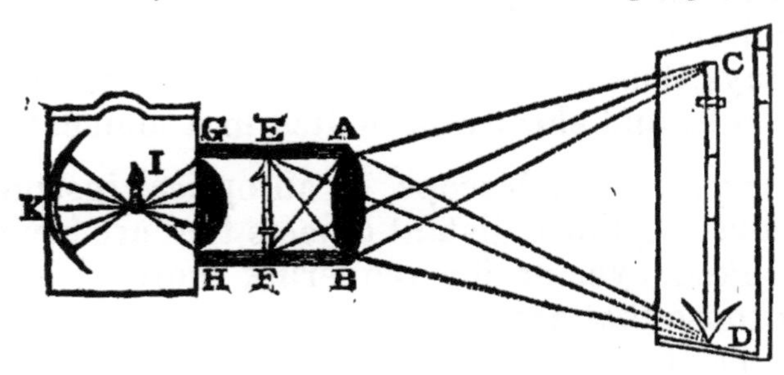

its side; G H is a thick plano-convex lens, which

strongly illuminates the object E F when it is put inverted into the tube; K is a concave reflecting mirror, to give additional force to the light; and A B is a double convex lens, placed in a movable tube, which slides in the interior of the projecting tube G B: when this lens is properly adjusted, it throws the shadow of the object large and upright against the side of the wall. The magnitude of the shadow C D is represented as much larger than the image E F, as the distance C A is greater than E A.

Burning glass. This is a double convex, or plano-convex lens, which collects the sun's rays upon its surface, and converges them into a point called the focus: when the rays are thus concentrated, they burn with great ardour, and will melt the densest metals.

As all those rays which fall upon the surface of a lens are collected in its focus, the effect will be in proportion to the difference between the surface of the lens and the surface of the focus; therefore, if a lens four inches in diameter collect the sun's rays at the distance of a foot from the glass, the image at the focus will not be more than one-tenth of an inch broad, so that the surface is more than sixteen hundred times less than that of the glass; therefore the sun's rays are so many times more dense at that point than on the surface. Burning glasses have been made three feet in diameter, and the rays, after passing through them, have been collected again by another lens placed parallel to the former, so as to converge them into a still smaller point at the focus. By the intense heat of the rays thus combined, gold has been fluxed in a few seconds, and sheet-iron melted in a moment.

Telescopes. A telescope is an optical instrument for discovering those distant objects that are invisible to the naked eye, or for rendering more clear and distinct those that are discernible. It is constructed to act either by refraction or reflection.

No invention in the mechanic arts has ever proved more useful and entertaining than the production of the telescope: its utility both by sea and land is too well known to need observation. With respect to the knowledge of the heavenly bodies, we owe much to the invention of the telescope; for without such assistance, the science of astronomy must have been far short of its present state.

The first invention is attributed to John Baptista Porta, a Neapolitan, about two centuries and a half ago; but Galileo soon afterwards greatly improved it, and by this means added considerably to the catalogue of fixed stars. Galileo's telescope passed on for many years without material alteration, till Gregory and Newton undertook the construction of telescopes, and brought them to a considerable degree of perfection, which has been completed by Herschel and others in the present day.

There are many kinds of telescopes; but as it would greatly exceed our plan to enter into a description of them all, it will be sufficient to describe some of the most material, such as the Astronomical Telescope, the Day or Land Telescope, the Newtonian and Gregorian. First, let it be premised, that the object-glass is that lens which is placed at the end of the tube nearest the object; the eye-glass is that which is nearest the eye; and when there are more lenses than one in the tube, beside the object-glass, they are called eye-glasses likewise.

The astronomical telescope consists of an object and eye-glass fitted into a long tube. The object-glass, which is a segment of a large sphere, is made either double convex or plano-convex; the eye-glass is double convex, formed from a segment of a small sphere; and these glasses are placed in the tube at the common distance of their foci.

Suppose rays of light issuing from every part of the object H I fall upon the object-glass A B; in passing through it, they will be refracted and converged into the foci E K, where the inverted image of the object will be formed; then if the eye-glass C D, which is of

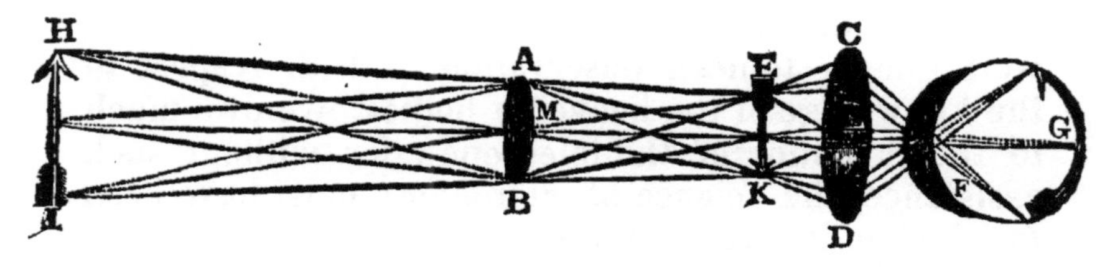

shorter focal distance, be so placed as to include E K, the rays will pass on through C D in a position nearly parallel, cross each other at F, and form a large, but inverted image of the object on the retina at G. The objects will be magnified by this glass in proportion as the distance of the focus of the object-glass M E exceeds the distance of the focus of the eye-glass E L.

The land telescope is used for viewing objects in the day time, on the surface of the earth. It is usually formed by three double convex eye-glasses, and a double convex or plano-convex object-glass. It exhibits the objects in an upright position, and the lenses are disposed in such a manner in the tube, that the distance between any two may be the aggregate of the distance of their foci; so that an eye placed in the focus of the first glass will see objects upright and distinct, and magnified in the ratio of the distance of the focus of the object-glass to the distance of the focus of the eye-glass at the opposite extremity.

If A B be the object, the rays from which are received by the object-glass C D, they enter the first

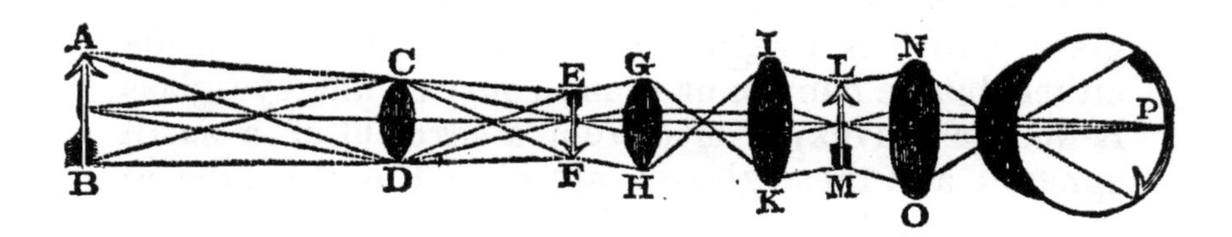

eye-glass G H; but instead of falling into the eye, as in the astronomical telescope, they pass on to I K, another lens equally convex, which is placed at double its focal distance from G H, so that the rays are transmitted parallel through the interval between them, and cross each other in the common focus of G H and I K.

After passing I K, the rays are again converged into the foci L M, where the image is formed in a position the reverse of E F; these rays are again transmitted through N O, and are then collected on the retina of the eye at P, where the image is clearly formed, and in an upright position.

As the addition of glasses in the land telescope does not magnify the object, an astronomical telescope may be used as a land telescope, by having an extra tube with eye-glasses made to slide into the end of the telescope; or that which is used in the day may be used for astronomical observations, by taking out two of the glasses.

Land telescopes are sometimes made with three glasses only, and some with five; but the dimness of the latter is equally inconvenient with the false representation of the former; so that four glasses, the medium, appears the best calculated to avoid the imperfections of either.

The aberration and colouring in the rays of light as they are transmitted through lenses, was a great obstacle to the improvement of telescopes, till a late optician, Dollond, contrived to form the lenses of different kinds of glass, which mutually correct each other's refrangibility, and greatly remedy the defects. An instrument thus fitted up, is called an Achromatic Telescope.

Application of the general law of reflection to the Telescope.

Reflecting telescopes are those which are chiefly formed by mirrors, and reflect the object to the eye, instead of refracting. They are principally confined to two kinds, called the Gregorian and Newtonian.

In the construction of the *Newtonian*, let A B C D be a large tube, open at the end C D, and closed at A B; the length B D being, at least, equal to the focus of the metallic reflector F E, which is placed near the end of the tube. The rays G *g*, I *i*, that come from H, a dis-

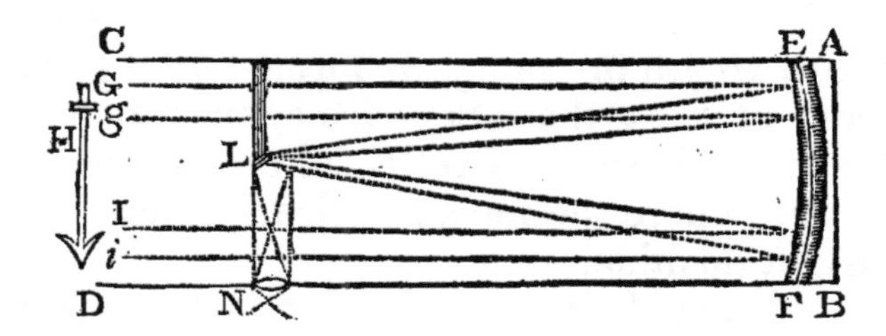

tant object, being considered as parallel to each other, fall on the concave speculum E F, and are reflected back upon a small plane speculum L, which is inclined in an angle of 45°; from this mirror they are again reflected to a convex lens, which is fixed in the side of the tube, and converged into N, the focus of the glass, where the image appears magnified and distinct to the eye. By fixing an additional tube, with lenses to the side of the telescope at N, the object may be either increased farther or diminished, and brought into an upright position.

The *Gregorian* telescope A B C D, is a large brass

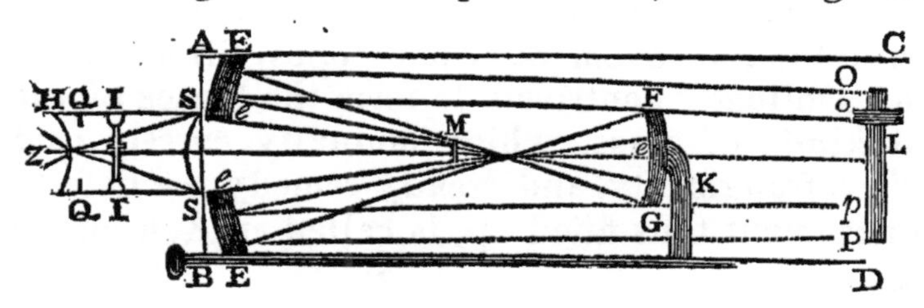

tube, in which is placed a concave metallic speculum E E, with a round hole *e e* perforated in the middle; F G is a small concave mirror fastened to the rod K, which is moved backwards or forwards at pleasure. If L represent an object at a considerable distance, and its rays O *o*, P *p*, enter the tube parallel to each other, they will fall on the larger speculum E E, from which they are reflected into the foci at M, where an inverted image is formed; but after crossing each other, the rays fall on the concave speculum F G, the centre of which *e* is the axis of the tube *z c*. From F G the rays would again converge into the foci Q Q, with image upright; but in converging to bring them close to the eye, they fall upon a convex lens at S S, by which the image is formed in the focus between I I, and thence taken up and carried to the eye at *z* by a meniscus H, where the image appears magnified, upright, clear, and distinct.

The latter of these reflecting telescopes is now generally used, as it shows all objects in their natural position, and is of a form the most convenient for portability and readiness in management.

Application of the laws of refraction and reflection to the phenomena called ***Rainbow, Halo,*** *and* ***Parhelion.***

Remark. The *rainbow* never appears, but when the sun shines on drops of rain, and is on the side of the spectator opposite to the centre of a circle, of which the bow in sight forms an arc. When a rainbow appears, the nearer to the horizon the sun is, the larger the rainbow will appear.

Large rays of light may be twice refracted and once reflected in drops of rain, producing an inner rainbow; others may be twice reflected and twice refracted, producing an outer rainbow.

Illustration. Numerous rays from the sun strike upon a drop of falling rain, and are reflected and refracted in all directions. Among others, suppose one minute pencil of rays to be refracted into the drop at A, and passing through to the inner surface of the opposite side at C. Most of the light would pass through it, but part would be reflected back to E; there it would be refracted, and then pass on to the eye of the spectator. Being refracted twice, the differently coloured rays would be separated, as in the prism. The violet being most refracted, would be highest; the red least, would be lowest; that is, the upper side of the bow, made up of such drops,

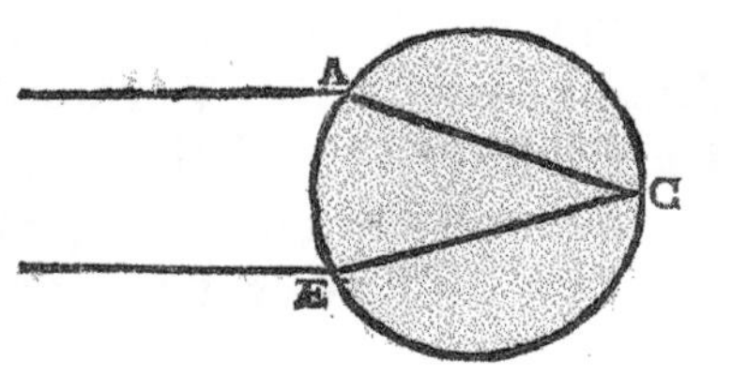

would be violet, the lower side red, &c. But the upper bow might have the violet below and red above, by two refractions in each drop of rain, at M and N, and two refractions at I and A; because the rays would be refracted upwards at I and downwards at A, bringing the most refracted rays downwards.

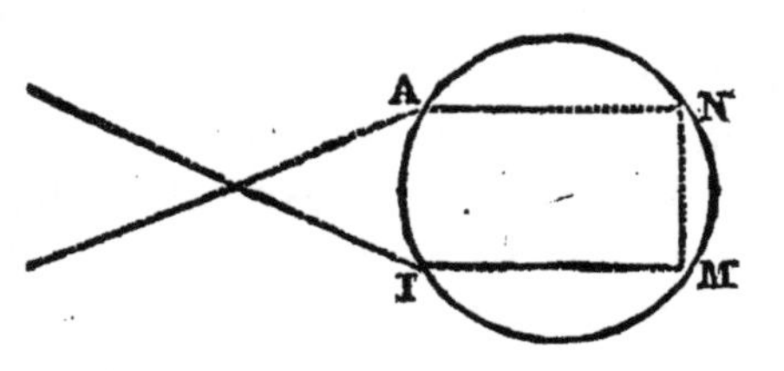

As every rainbow is formed by the sun's rays upon drops of rain, the preceding description is probably correct. As more light is lost at these feeble reflections than at the refractions, the upper bow, with its two reflections, is always the most feeble.

Halos, or Coronas, are circles about the sun, moon, &c. made by refraction or reflection of light in visible aqueous vapour.

Illustration. Fill a room with steam, by pouring water on hot iron. A candle will present the same phenomenon, at a distance from its centre equal to the refracting or reflecting angle of equally dense steam in all cases.

Parhelia, or Mock Suns, are probably produced by reflection from dense vapour.

Illustration. The fore part of November, in the year 1820, when the sun was about half an hour high at evening, I called the attention of a part of the medical class in the Vermont Academy of Medicine at Castleton, to the following phenomena:

A luminous circle appeared round the sun, excepting that the lower part was hidden beneath the horizon. On each side of the sun, its face seemed to be reflected, with a train on the side of each opposite to the sun, as shewn in the figure. A much larger circle was formed above the sun, with two luminous spots also, but much fainter. Above this was another

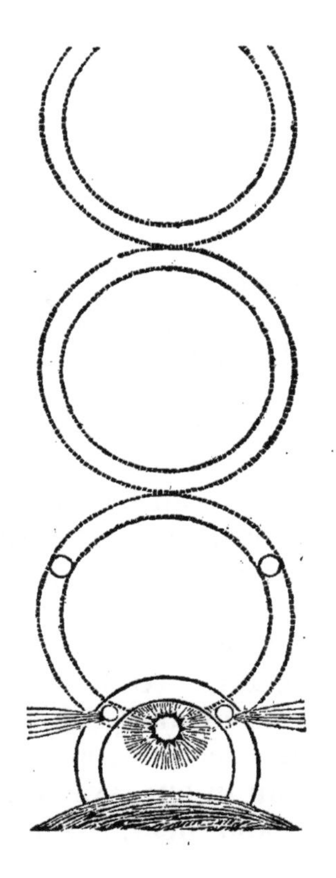

circle, and still above the last almost another, as in the figure. All these circles and spots seemed to be reflected upon dense brownish yellow vapor, which covered the sky from the zenith to the horizon. The face of the sun could be seen, but the vapour so far obscured it, that the eye was not pained by looking directly upon it.

I attempted to illustrate it to my pupils at the time, by imagining the sky the inside of a concave mirror. Though I was not perfectly satisfied with my own explanations, it appeared most reasonable to suppose them produced by reflection; for many cases are on record of the reflection of ships, &c. in the aqueous vapour above them. The law expressed by the angles of incidence and reflection would locate these circles very nearly, or perhaps precisely, as they appeared, if the mirror is sufficiently concave.

X

SOLAR SYSTEM.

Remarks. The power of God being made manifest even in the smallest of his works, how much must the human mind be led to contemplate his infinite wisdom and power, when we survey the multitude of stars that are scattered through the infinity of space! Judging from analogy of the general purposes of creation, we can hardly conceive them to be placed for mere ornament, or even for the purpose of giving light in the absence of the sun, when the reflection of his rays from the moon, a single satellite, gives a thousand times more light to the earth than the whole of the stars. Rather let it be presumed, that the great number of stars, which have no revolutionary motion, are destined for more important purposes; and, like the sun in our system, are inexhaustible fountains of light and heat, which diffuse their vivifying powers to their own surrounding orbs; the opacity of whose bodies, and the immensity of their distance from us, render them invisible to our eyes. Thus the extent of imagination falls infinitely short in comprehending the greatness of God, or his works. We form a comparative idea of the distance of a mile, a thousand or a million of miles; but where are the bounds of that almighty power, which created those innumerable systems that are scattered at such immense distances from each other through the infinity of space?

The sun, and the planetary worlds which revolve round it, is one of those numerous systems that we are now about to consider.

Various opinions have been adopted, at different times, with respect to the motion of the sun and planets; but as the Copernican system is now established, and accepted by all enlightened nations as the most consonant to reason and the operations of nature, it will be sufficient for our purpose, before we enter into an explanation of it, merely to mention two others, which at different times have had their disciples.

The first is called the Ptolemean, from Ptolemy, its framer, who was born in Pelusium, in Egypt, and flourished as a great mathematician and astronomer soon after the commencement of the Christian æra.

Guided by the sensible appearances of the heavenly bodies, without considering either their absolute or relative motion, he considered the earth as a stationary body, fixed in the centre of the system, and that the sun and planets were subordinate, and revolved round the earth in twenty-four hours.

After the Ptolemean, the other was formed by Tycho Brahe, a Dane, who considered the earth to be placed in the centre of the universe, and that the sun revolved round it, whilst the rest of the planets revolved round the sun. In an improvement of this system, a diurnal motion was given to the earth round its own axis, to account for day and night more naturally than by a revolution of the whole system in twenty-four hours. But this complicated and ill-digested hypothesis soon fell into disrepute, to make way for that called the Copernican system; which is long likely to endure, as a monument of human ingenuity, and a rational system of the planetary motions.

Nicholas Copernicus was born at Thorn, in Prussia, in the year 1473. He rather revised and perfected the doctrine of Pythagoras, who existed about 600 years before Christ, than created any new system of his own. The Pythagorean idea of the universe attracted the mind of Copernicus, and after a labour of twenty years, he brought it to perfection, and died just in time to save himself from the bigoted persecution of the Romish church for his discovery.

*Description of the Solar System.**

The sun is placed in the centre of our system, and the earth and the other planets revolve round it in different orbits, at great distances from each other.

These planets, which we perceive by the reflection of the sun's rays from their opake bodies, are of three kinds, called primary, secondary and comets.

The primary planets are those which move round the sun in orbits nearly circular and concentric, at different distances from it. They are seven in number, called ☿ Mercury, ♀ Venus, ⊕ Earth, ♂ Mars, ♃ Jupiter, ♄ Saturn, and ♅ Herschel. Mars, Jupiter, Saturn and Herschel, are usually called superior planets, because their orbits include that of the earth. Venus and Mercury are called inferior planets, as their orbits are contained within the earth's orbit.

The secondary planets, satellites or moons, are attendants on primary planets, and revolve round them. The earth is accompanied by one moon, Jupiter by four, Saturn by seven, and Herschel by six.

The comets revolve round the sun in very eccentric orbits. Their number and periodical revolutions have not yet been correctly determined: they move in very eccentric orbits, suddenly appear and disappear, and are usually attended by a long train of light, which is called the tail of the comet.

The adjoining diagram is the usual representation of the solar system in plano; but it is not strictly correct, as the planets move in elliptical orbits, the planes of which do not exactly coincide with one another.

* It is not convenient to divide an abridged system of astronomy into principles and illustrations, because experiments and familiar references cannot generally be introduced. The most important elementary facts, however, are printed in italics

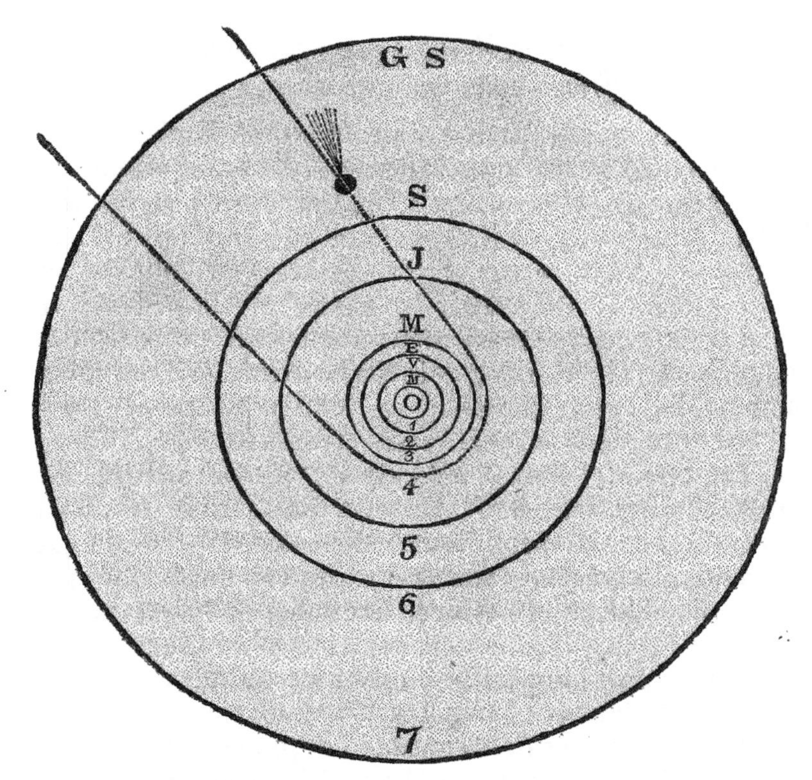

O represents the sun placed in the centre of the system, M 1 is the orbit of Mercury round the sun, V 2 the path of Venus, E 3 the orbit of the earth, M 4 the orbit of Mars, J 5 the course of Jupiter, S 6 that of Saturn, and G S 7 is the orbit of Herschel.

All the planets are supposed to have a compound motion, like the earth, called the annual and diurnal.

The annual motion is that with which they pass through their orbits from west to east, forming a year by one complete revolution round the sun.

The diurnal motion is the rotation of a planet round an imaginary line passing through its centre, called its axis, whilst it is moving through its annual orbit: one complete rotation is called a day. This compound planetary motion may be conceived by the rolling of a ball upon a table, which is perpetually turning round whilst it passes from one extremity to the other.

All bodies at great distances appear equally distant, though the difference is very great.

The sun, moon, planets, and fixed stars, all appear to be placed in the same concave sphere, of which the eye of the spectator seems to be the centre; so that the bodies apparently differ in magnitude, but not in their distances. We estimate the distance of objects on the surface of the earth by some given measure or comparative proportion with some other objects less remote; but in viewing celestial bodies, the immensity of their distance affords us no relative means of forming our judgment of their respective positions. Therefore the optical sense is deceived; for demonstrations show us that the sun is nearer to us than the fixed stars, the ocular proof from eclipses convinces us that the moon is nearer the earth than the sun, and our reason teaches us to believe that some of the stars are many millions of miles nearer to us than others. But an easy experiment will show how unable we are to judge of distances by our sight alone; for, in a dark night, if a few lighted candles be placed at different distances from a spectator, they will all appear equally remote, but the flames will vary in magnitude according to the distance, and the person will be unable to judge, with any correctness, how far he is from them.

The planets in their annual courses cross the ecliptic, or sun's apparent path, in two opposite points called the nodes. But the planets do not move in the same plane with each other, as they cross the ecliptic in different parts of the heavens. This may be properly represented by placing different sized hoops within each other in different directions, considering the centre as the sun's place, and the hoops themselves as the orbits of the planets.

The orbits of the planets are elliptical, with different eccentricities.

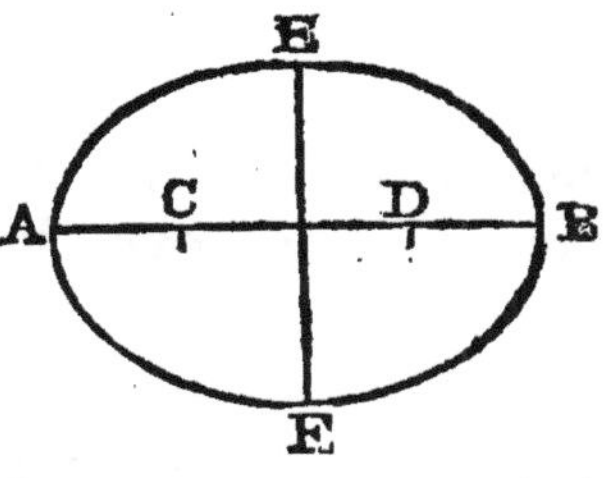

A E B F represents the orbit of a planet, with the sun in one of the foci C D; the axis A B is called the line of the apsides; and when the sun is at D, and the planet at A, its greatest distance from it, the planet is said to be in its aphelion, or higher apsis; when it is at the other extremity B, or nearest the sun, it is then in its perihelion, or lower apsis.

The mean distance of a planet from the sun, is between the extremities of the conjugate diameter E F, *and of the transverse diameter* A B.

Two planets are said to be in conjunction, when they both appear equally advanced in the same part of the heavens; and when they are in opposite points, they are said to be in opposition.

Each planet has its peculiar course, which it always pursues without deviation. The whole courses of the planets are included in a certain zone or belt of the heavens, extending between 18° and 19° in breadth, which is called the Zodiac, containing the constellations Aries, Taurus, Gemini, Cancer, Leo, Virgo, Libra, Scorpio, Sagittarius, Capricornus, Aquarius, and Pisces.

The whole solar system is supposed to revolve around a general centre in the universe. Because the stars in one part of the heavens seem to approach each other, and to recede from each other in opposite directions. This is evidence that the solar system is in motion, in reference to the stars. It is hence inferred that this motion is circular, because we are acquainted with no rectilinear motion among the heavenly bodies.

Description of the Sun and each Planet.

The Sun is the great luminary which dispenses light and heat to all the planetary system. It has been usually reckoned amongst the planets, but it more properly belongs to the fixed stars, as one of those central bodies dispersed through the infinity of space which have their subordinate orbs revolving round them. The sun is placed nearly in the centre of our system, and revolves round its own axis in 25¼ days: the axis has an inclination of about eight degrees with the ecliptic.

Although its apparent diameter is seen from the earth under an angle of 32′ 12″ only, its real diameter is not less than 890 thousand miles, and it is about 1,392,500 times bigger than our earth, which is 95 millions of miles distant from it. It appears to us to have a revolving motion through an orbit from east to west; but this apparent motion will be hereafter shown to arise from the diurnal motion of the earth from west to east, whilst it is passing through its annual orbit.

Though the sun's atmosphere is luminous, its body is probably dense and opake, like the planets. When the sun is viewed through a telescope, it seems to have dark spots on its disk. These spots appear to be notches when they come to the periphery of the sun. Hence it is inferred that they are holes or openings in its atmosphere, exhibiting the opake body of the sun.

Mercury is the least of all the planets, and nearest the sun, which makes it seldom visible to us, as its reflected light is lost in the sun's more powerful rays. Its greatest elongation, or apparent distance from the sun as viewed from the earth, is not more than 28°. It is computed to be about 37 millions of miles distant from the sun, and revolves round its orbit in 87d. 23h. which forms its year. The diameter of Mercury is three thousand miles; it contains 28,274,000 square miles on its surface, and moves at the rate of 110,680 miles in an hour. When it is seen through a telescope, its edge appears clear and distinct. Its body is opake, and, like the moon, reflects a borrowed light, and changes its phases or appearance according to its several positions. When it passes over the sun's face, or is between us and the sun, this is called its transit, and the planet appears like a black spot in the sun's disk.

Venus has generally a larger and brighter appearance than any other planet, which makes it easily distinguishable from the rest.

Its diameter is 7699 miles, and its distance is 69,500,000 miles from the sun : it revolves through its orbit, or completes its year in 224d. 6h. and moves at the rate of 80,955 miles in an hour. Venus forms its day, or turns round on its own axis in 23h. 22m. and its greatest elongation from the sun is about 48°. Like Mercury, it is invisible at midnight, and is only seen for two or three hours in the morning or evening, when it passes before or after the sun.

The Earth is placed next to Venus in the planetary sphere; its diameter is 7920 miles, and it is about 95 millions of miles distant from the sun. It makes one complete revolution in its orbit in 365d. 5h. 48m. moving at the rate of 68,856 miles in an hour. The earth turns round its own axis from west to east in 23h. 56m. which produces the apparent diurnal motion of the sun and all the heavenly bodies from east to west in the same time : the diurnal motion of the earth likewise causes what we call the rising and setting of the sun, and the length of days and nights. The axis of the earth is inclined 23½° to the plane of its orbit ; and as this axis is always parallel to itself, or in the same direction in every part of its course, it causes the sun at one time of the year to enlighten more of the northern parts of the globe, and at another time of the southern, which produces the various seasons of spring, summer, autumn, and winter.

The Moon is a secondary planet, and an attendant of the earth, revolving in an elliptical orbit, or rather the earth and the moon both revolve round a common centre of gravity, which imaginary point is as much nearer to the earth as the mass of the earth exceeds that of the moon. The moon makes its revolution in its orbit round the earth in 27d. 7h. moving at the rate of 2299 miles in an hour. Its time in going round the earth, reckoning from one new moon to another, or when it overtakes the sun again, is 29d. 12h. It is 2161 miles in diameter, and 240,000 miles distant from the earth, turning round its own axis in the same time that it revolves round the earth, so that its days and nights are of the same length as our lunar months. The moon's orbit is inclined to the plane of the ecliptic in an angle of about 5°, and crosses it in two opposite points called the nodes: lunar eclipses take place when the moon is in or near these points.

Mars is the first of the superior planets, being placed on the outside of the earth's orbit: it is 5309 miles in diameter, and its distance from the sun is about 146 millions of miles; it performs its revolution round the sun in 1y. 321d. 23h. moving at the rate of 55,287 miles in an hour, and revolves round its own axis in 24h. 39m. This planet has a greater analogy to the earth than any other planet: the diurnal motion and the obliquity of its ecliptic have very small variation from those of the earth. When it is in opposition to the sun, it is five times as near to us as when it is in conjunction, which has a very visible effect on its magnitude. It has a dusky and reddish hue, which is supposed to arise from the nature of the atmosphere that surrounds it.

Jupiter is placed between Mars and Saturn, 90,228 miles in diameter, and it is about a thousand times bigger than the earth: its distance from the sun is 499,750,000 miles, and it revolves round its own orbit in 11y. 314d. 12h. moving at the rate of 29,000 miles in an hour. It has a diurnal motion round its axis in 9h. 56m. which carries the equatorial parts of its surface with a velocity of 25,000 miles an hour: this is about twenty-five times faster than the revolution of the same parts of the earth. Jupiter has four satellites revolving round it, which enlighten it in the absence of the sun, as the moon enlightens the earth. Beside these attendants, it is surrounded by faint bodies, which are called its zones or belts: these appearances are frequently changing, and are ascribed to clouds in its atmosphere. As the axis of Jupiter is nearly perpendicular to its orbit, there is hardly any difference in the seasons, and the days and nights are five hours each.

Saturn has been considered for many ages as the last and most remote planet in our system, until some recent discoveries. In consequence of Saturn's great distance from the earth, it casts but a feeble light of a dusky colour, although it surpasses all the rest, Jupiter excepted, in actual magnitude. Its diameter is computed to be 79,979 miles, and its distance from the sun 916,500,000 miles, which is near ten times the distance that the earth is from the same luminary. It takes 29y. 167d. to make one complete revolution in its orbit, and moves at the rate of 22,298 miles in an hour. This planet is surrounded by two rings, one within the other; and beyond these rings are seven attendant moons, two of which were discovered by Herschel.

Herschel was discovered by Herschel in the year 1781: its light is of a bluish white colour, and may sometimes be seen by the naked eye in a clear night without moonlight. The time of its annual revolution is about 80 years, and its diameter 34,299 miles, which is more than four times the diameter of the earth. Its distance from the sun is about 1832 millions of miles, and it has an inclination of 43° 35′ to its orbit.

To assist the memory, and form an idea of the proportional distance of each planet from the sun; if the greatest extent of Herschel from the sun were divided into 190 parts, the proportional distance of the rest of the orbits would be, Mercury 5, Venus 7, the Earth 10, Mars 15, Jupiter 52, and Saturn 95.

Besides the primary and secondary planets and comets, other bodies have recently been discovered, which resemble these in their motions, &c. Four small bodies revolve between Mars and Jupiter, called Juno, Ceres, Pallas, and Vesta. They seem to be intermedial between planets and comets. Their orbits are more elliptical than those of planets, and less than those of comets. Three of them have trains somewhat like those of comets.

Another kind of body, called Aerolites, are now generally considered as small revolving bodies, somewhat like satellites, whose orbits are not far distant from the outer limits of our atmosphere. When the centripetal force is greatly increased by the sun and larger planets being on the opposite side of the earth, they are supposed to be drawn within the limits of the atmosphere ; atmospheric friction and resistance causes ignition, and finally their fall to the earth.

Geometrical Trigonometry applied to Astronomical Calculations.

Suppose the distance from the centre of the earth to the centre of the moon is required, make a triangle as follows: The moon in the sensible horizon at c is in one angle, the centre of the earth at *e* another, and the place of observation on the surface of the earth at *a* the other. *a* is a right angle; *a e* is the known semi-diameter of the earth. This exhibits what is called the moon's horizontal parallax. Now if we had the angle *c* at the moon, we could construct the triangle, and measure the line *e c* by the scale by which the line *a e* was laid down, which would be the distance from the centre of the earth to the centre of the moon.

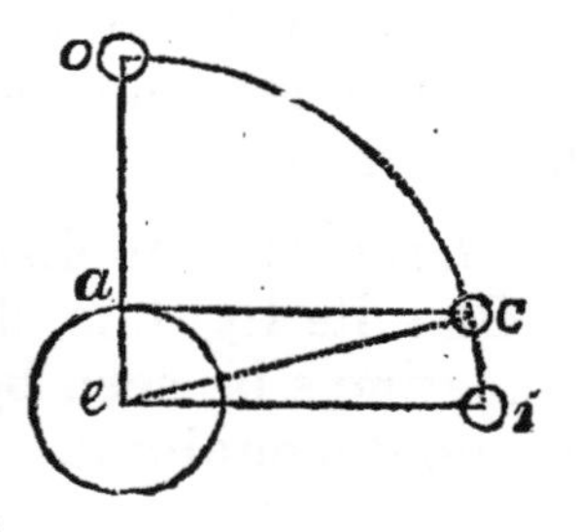

To find the angle at *c*, we must first find the angle *c e i;* then the line *e* c being a diagonal, the angles at *c* and at *e* are equal. The only remaining object being to find the acute angle at *e*, we must proceed as follows: Find the time required for the moon to perform its apparent revolution round the earth: one-fourth of this time is required for its apparent progress from *o* to *i*, the rational horizon. By accurately measuring the time, it will be found that the moon arrives a little too soon at the sensible horizon at *c*, after making the usual allowance for refraction, &c. Whatever this difference may be, say, as the time of the whole revolution is to the time of its difference, so is 360 degrees to the arc *c i*, the measure of the angle at *e*. This gives the angle at *e*, which is equal to the angle at *c*, as required for perfecting the triangle *a c e*.

This method of calculating the distance of heavenly bodies is sufficient for the short distance of 240 thousand miles, the moon's distance from the earth; but the angle at the body of the sun and planets is too small for the most accurate instruments. The passage of Venus over the sun's disk was therefore used by astronomers for settling their distances more accurately.

Method of finding the sun's distance from the earth by the transit of Venus, as explained by Prof. Douglass, of West Point, to the cadets, May 25, 1822:

First, find the relative proportions of the distance from the earth to the sun, and from Venus to the sun. This is done by constructing a right-angled triangle,

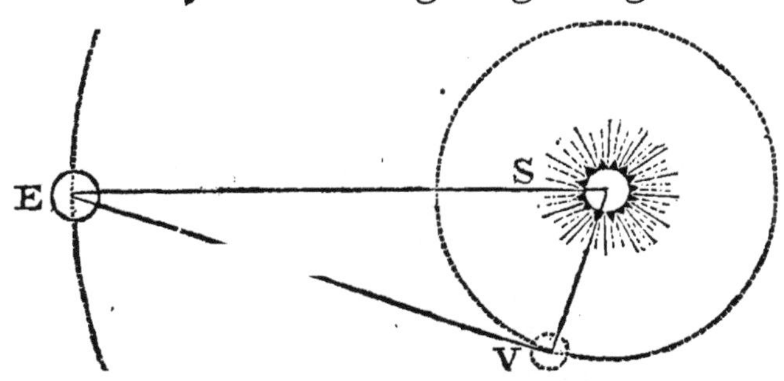

with its right angle at Venus V, one acute angle at the earth E made by a line drawn from the earth to the sun S, and another line from the earth to Venus when it is at its greatest apparent distance from the sun: the angle at the earth may be taken by the quadrant. Now, suppose any length for the hypothenuse, (or distance from the earth to the sun,) say one hundred million of miles: close the triangle geometrically, and measure off the distance by the scale; or say arithmetically, as the sine of 90° at Venus is to one hundred million miles, so is the sine of the angle at the earth to the distance from Venus to the sun.

Now, if we can find the true length of the side from Venus to the earth, we can, by this proportion, find the distance from the earth to the sun, the sun to Venus, &c.; and these distances will lead to the distances of all other planets, by considering the times of their revolutions, &c.

Make a triangle A B and ☉, with the side A B equal to the hypothenuse of the preceding figure, and the

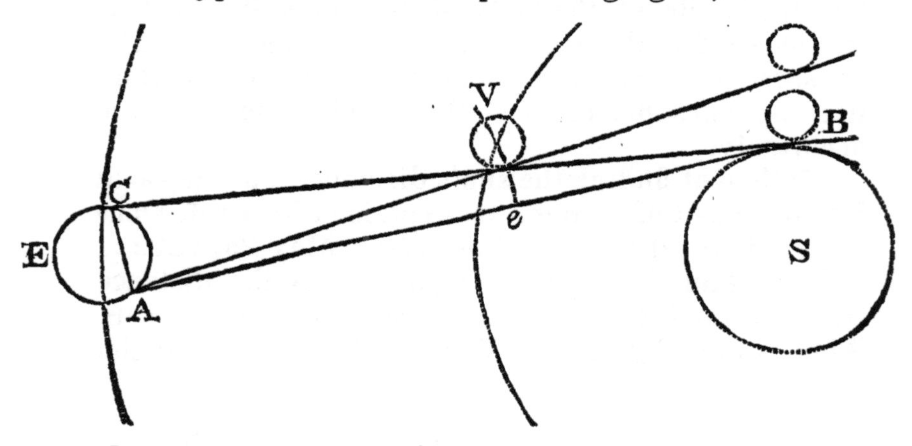

side B ☉ equal to the side from Venus to the sun; complete the triangle by the side ☉ A: then make the triangle A C ☉, by making A C equal to the ascertained chord line between the stations. Now, if the angle at A can be found, all the angles can be found; because, though we have not the true measure of any of the sides, yet having the relative proportions of two sides, we can find the proportions of all three of the sides, and the true measure of all the angles. All this will be exhibited by a geometrical diagram, after the angle ☉ A *e* is known and accurately laid down.

But if we calculate arithmetically, we can say, as the side B O is to the angle at A (when we find it,) so is the side A B to the angle at O : then we can find the proportion of the side A O also, &c.

The angle at A is found by the transit of Venus thus : One astronomer stands at C, and another at A. The observer at C sees the limb of Venus touch the sun at 5 o'clock P. M.; the observer at A sees the same 10 minutes later, after allowing for his being perhaps 120 degrees farther west : during that time, Venus has apparently travelled from O to *e*; that is, described the arc subtending the angle at A. To know how large that makes the angle, they continue their observations until Venus crosses the disk of the sun : suppose this requires 40 minutes, then the arc O *e* is one-fourth of the arc across the sun. The disk of the sun is known to subtend an angle of 32⅕ minutes on the earth ; then the angle at A would be one-fourth as much, or 8 minutes and 3 seconds. With this, all the angles, and the proportions of the sides of the triangle A B O, can readily be found, as before mentioned.

Now, all we want is the true measure of one of these sides, in order to settle the true measure of all the sides ; and we can easily find the true length of the side A O : for it is also a side of the triangle A C O, and in this the angles at A and C are taken by the observers with the quadrant and plumb-line, or artificial horizon, sufficiently accurate for deciding between them. Or, we may suppose one of the angles to be a right angle : the acute angle at O may be found by subtracting half the obtuse angle at O from 90°, and the remainder will be the acute angle at O required. To obtain this angle, as before directed, with perfect accuracy, is the chief use of the transit. Then say, as the acute angle at O is to the side A C (the ascertained chord line between the observers,) so is the angle at C to the side A O. Having the true measure of the side A O, and all the angles in the triangle A B O, the other two sides are found by the common rule. Therefore we have A B the earth's distance from the sun, O B the distance of Venus from the sun, and O A the distance of Venus from the earth.

Three corroborating calculations may be made on the same principle, founded on observations taken at the same time, viz: When the upper edge of Venus enters upon the sun; when the lower edge comes out; and when the upper edge comes out.

The Earth and Moon considered, with an explanation of Seasons and Eclipses.

In the early ages, the opinions of mankind were much divided concerning the form of the earth; some being guided by visual appearance, conceived it to be a stationary plane, bounded by the horizon, and that the whole universe was contained in that part of the heavens which was presented to their view. However ill conceived this opinion may seem in the present day, it has had its supporters, even in the ages of Christianity. But as a strong proof that all the sages of antiquity were not equally ignorant of its real form, we find the ancient Babylonians and Greeks calculated eclipses both of the sun and the moon, which may be taken as a fair argument to show that they were not unacquainted with the rotundity of the earth.

The earth is of a globular form. The number of convincing proofs which are produced in the present day, cannot leave a doubt of its globular form, even in the commonest minds; for those who stand on the sea-shore and observe a ship making out for sea, will perceive the hull first decline as it approaches the horizon, till it totally disappears, leaving the mast and sails still in sight, and these gradually decline till the top of the mast sinks from the eye. Even then, if the spectator ascends the top of a hill or building, he will perceive the vessel again, till the convexity again hides it from his sight.

It is perfectly clear, that if the earth were a plane, the hull of the vessel, which is the largest part of the body, would be seen the longest, and the mast and sails would first disappear, as the inferior objects of sight; but observation proves the reverse.

When a ship is sailing at sea, either northwards or southwards, those stars which are placed nearly opposite to the poles of the earth, appear to have no diurnal motion, but remain fixed in the extreme parts of the heavens; therefore, if the earth were a plane, considering the immense distance of the stars, a ship in sailing either directly north or south would still observe them under the same angle, or with the same altitude; but daily experience teaches us the contrary, for vessels sailing northwards observe a gradual elevation of those stars which are in the north polar regions, and a depression of those towards the south: in sailing southwards, the appearance is the reverse, for the southern stars are elevated and the others depressed. This appearance is rationally explained by the convexity of the earth, which increases the angle of observation as the ship sails towards the star, and decreases it as the vessel moves the opposite way.

Eclipses of the moon are caused by the earth's shadow falling upon it, when the earth's body is interposed between the sun and the moon; yet we always find, that in whatever position the earth may be placed at that time, its shadow falls with a circular edge upon the disk of the moon, which could not always happen if the earth were not of a globular form.

The irregularities on the surface of the earth have no visible effect upon its shadow on the moon; for the highest mountain on its surface, considering the magnitude of the earth and its distance from the moon, would cause no more visible effect in its shadow, than the finest grain of sand would produce on that of a billiard ball.

The globular form of the earth is likewise practically known by those circumnavigators who have sailed around it, by always continuing an easterly or westerly course, which has brought them again to the same port whence they set out.

The earth is a little flattened at the poles.

These and various other proofs in the higher departments of science, leave no doubt of the sphericity of the earth, although it differs, in some measure, from a sphere, as it is flatted towards its poles, something resembling an orange; so that the diameter of the earth at the equator exceeds that at the poles by about thirty-four miles. This was discovered by observing that a pendulum moved slower as it approached the equator, and faster as it advanced towards the poles. This difference was probably caused by the centrifugal motion of the earth on its axis, which diminished the force of gravity towards the equatorial parts of the globe, and flattened the earth towards its polar extremities, while the earth was in a semi-indurated state soon after its creation.

The earth has a diurnal and annual motion, dividing time into days and years.

Independently of a small motion which occasions what is called the precession of the equinoxes, the earth has two general motions; one round its own axis in twenty-four hours, which is called its diurnal motion, and causes the succession of day and night; the other is its annual motion or revolution round the the sun as its centre, keeping its axis always inclined to its path in an angle of about 23½ degrees, which produces the various seasons of spring, summer, autumn, and winter.

To the visual sense it would seem that the earth is fixed as the centre of our planetary system, and that the sun and the rest of the celestial bodies have a daily motion round it. It is not extraordinary that men in early ages should have considered this as the system of planetary motion, when at the present hour, the uninformed, who judge only from sight, will not be persuaded to give up their opinions; but to those who are susceptible of conviction, there are so many proofs of this error, that nothing but ignorance or obstinacy can hesitate to believe them.

The following observation may tend to show relative motion, and how easily our senses may be deceived.

If a person be placed under the deck of a vessel when it is sailing gently down the side of a coast with a fair wind, he will be perfectly insensible of its motion; but if he cast his eyes on the shore, he will see all the objects pass him with a rapidity equal to the velocity of the vessel, and the vessel itself will be apparently at rest. But here our reason tells us that our senses are deceived, and that the motion is in the ship and not in the objects. Then why cannot we suffer ourselves to believe that our sight is deceived by the apparent motion of the sun, and simplify the system of the universe by admitting a diurnal revolution of the earth from west to east, rather than force such a monstrous hypothesis as would drive the whole universe round the earth from east to west to form the period of a day.

By the diurnal motion of the earth, the same phenomena appear as if all the celestial bodies turned round it; so that in its rotation from west to east, when the sun or a star just appears on the eastern side of the horizon, it is said to be rising, and as the earth continues its revolution, it seems gradually to ascend till it has reached the meridian, which is directly south of the observer; here the object has its greatest elevation, and begins to decline till it set, or become invisible on the western side. In the same manner, the sun appears to rise and run his course to the western horizon, where he disappears, and night ensues, till he again illuminate the same part of the earth in another diurnal revolution.

All the heavenly bodies do not appear to rise and set: those that are placed in the two poles, or are opposite to the imaginary axis of the earth, can have no apparent motion from its daily revolution, therefore always appear in the same part of the heavens.

In summing up the diurnal motion of the earth, it is considered as a globular opake body, turning round on an imaginary axis from west to east, and that it is enlightened by the sun's rays, which perpetually illuminate one half of its surface: the imaginary great circle which separates the illuminated part from that which is dark, or turned from the sun's rays, is called the terminator; and when any point on the eastern side of the globe comes to this line, it is sunrise; and when the point advances to the edge of the terminator on the opposite or western side, it is called sunset. The light which gradually appears before the rising of the sun, and gradually decreases after sunset, is called the crepusculum, or twilight, and is occasioned by the reflection of the sun's rays from the atmosphere that surrounds the earth.

Before we speak of the annual motion of the sun, it may be seen by the following diagram, that if the earth be at rest and the sun in motion, or if the sun be at rest and the earth in motion, the same general appearance and effect will be produced; that is, either of them would apparently describe a great circle in the heavens in the plane of the ecliptic.

For, if the earth be supposed at rest, and the sun revolving round it in the orbit A B C D, it would appear to a spectator on the earth to describe the great circle F E G H in the heavens; for if a spectator on the earth at O view the sun at A, it will be referred to E, and at B to F, and so on of the rest.

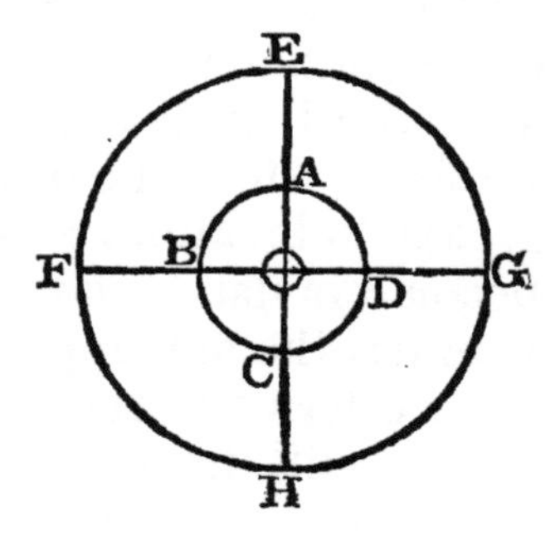

But if the sun be placed at O, and the earth in the orbit A B C D, when the earth is at A, the sun will appear in a great circle of the heavens at H; when it is at B, the sun will be seen at G, and so on with the other two points.

Therefore, whatever regards the sun's place, with respect to its appearance in the heavens, it may be considered, instead of the earth, as moving in an infinitely great circle, called the ecliptic, having its centre in the eye of the observer.

Changes of the seasons are caused by the oblique position of the earth's axis in relation to the plane of its orbit.

The earth is supposed to be divided into two equal parts, by a great circle drawn at equal distances from the poles, which is called the equator: smaller circles drawn parallel to the equator, approaching the poles, are called parallels of latitude; and great circles intersecting the equator at right angles, and passing through the poles, are called meridians of longitude. The ecliptic is the earth's orbit, or the apparent annual course of the sun, making an angle of about 23½ degrees with the equator, from which the various seasons are derived; and the terminator is a great circle which bounds the illuminated part of the earth, or that half of its surface which is always turned towards the sun.

Let A B C D represent the earth, with its poles A B so placed that the terminator of the sun's rays passes through each pole; then it will likewise divide every circle or parallel of latitude *a*, *b*, *c*, &c. into two equal parts, so that one half will be enlightened by the sun's rays, and the other will be dark, or turned from them; but during the diurnal motion, every part of each circle, or each meridian, will be brought to the terminator, and carried through the illuminated part in the same time, which makes the days and nights of equal length on every part of the globe, except at the poles A and B, where the sun would be seen during the whole day just skimming above the horizon.

But if the axis of the earth A B do not coincide with the great circle of the terminator E F; the great circle C D, called the equator, will be divided into two equal

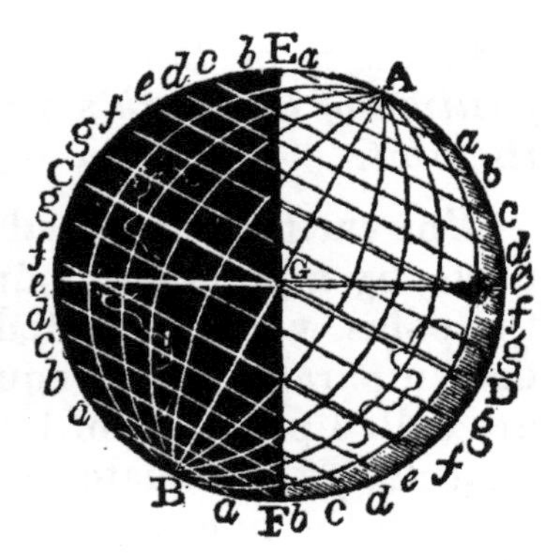

parts at G, and the inhabitants under that line will have their days and nights of equal length; but the parallels *a a*, *b b*, *c c*, &c. will be divided into unequal parts by the terminator E F, and the inhabitants under those parallels which have the greatest part of their circle illuminated, will have their days longer, and nights shorter, than those which lie on the opposite side of the equator, where the dark and illuminated parts are just the reverse. It is likewise clear from inspection, that as the diurnal and nocturnal parts of these parallels respectively increase from the equator, those places that lie under them will have a greater disproportion of day and night; so that those that are at or near the pole A will have constant day, or the sun always above the horizon; and those at the opposite pole B will have continual darkness, or the sun always beneath the horizon for six months together.

In the present position of the axis, those who lie on the northern side of the equator have their summer, and those on the southern their winter.

The difference of light is not the only cause of summer and winter. The sun appears much higher above the horizon to those places which have the longest days; consequently the rays fall more perpendicularly upon the earth, which, joined to the superior quantity of heat that is communicated by the greater length of the day, causes the summer to be much hotter than winter.

Having shown that the length of day and night is produced by the different relations of the axis of the earth to the terminator, we will next explain how this varies at different times of the year, and produces the variety of seasons.

We find, by common observation, that the sun describes different parallels, or daily appears at different heights above the horizon, which shows that the plane of the ecliptic, or sun's apparent path, does not coincide with the plane of the equator, otherwise the sun would have the same altitude daily, and days and nights would always be of an equal length. Let it be remembered, that the effect is the same, whether the motion be in the earth or the sun ; therefore, we more properly conclude, that the axis of the earth is inclined to its orbit, or that the plane of the equator does not coincide with the plane of the ecliptic. Now it is found by observation, that this inclination of the planes forms an angle of 23½ degrees, which remains invariable throughout the whole of the earth's annual course, or that the axis of the earth always moves parallel to itself.

Therefore, if the earth's axis be inclined to the plane of the ecliptic, as in the last figure, one of its poles will be towards the sun, and the other will be removed from it in an equal proportion ; but when the earth has made half of its revolution, and has arrived in the opposite point of its orbit, still retaining the inclination of its axis, the pole which was towards the sun will now be removed from it, and the opposite pole, which was darkened, will be presented to it. As the earth passes from one of these extremes to the other, the plane of the ecliptic will coincide with the plane of the equator, near the intermediate distance between the extreme points, and then the terminator passes through the two poles of the earth, as in figure the first.

From this revolution the seasons are produced ; for when the northern pole is the most inclined towards the sun, it is midsummer to all the inhabitants on the north side of the equator, who then have their longest day and shortest night ; but when the earth is in the opposite part of the ecliptic, or 180° distant, then the southern pole has its greatest inclination to the sun, and the inhabitants on that side of the equator have their midsummer and longest days, while those in the northern hemisphere have their shortest days and midwinter. When the earth is in the intermediate part of its orbit, or 90° distant from either extreme, that is, when the plane of the ecliptic coincides with the equator, the days and nights are of equal length all over the world, which is in the vernal and autumnal equinoxes, or spring and autumn.

The following experiments will show the effects which are produced by the inclination of the earth's axis to its orbit through all the twelve signs of the zodiac.

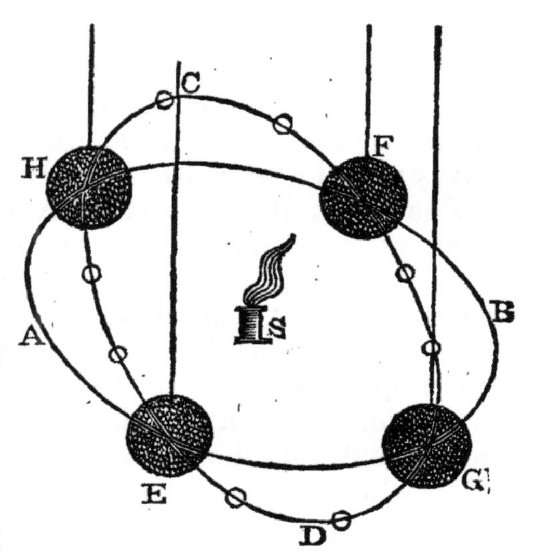

Let the frame C D represent the elliptical orbit of the earth, which intersects the equator A B in the two points or nodes E F, making an angle of $23\frac{1}{2}$°, which is the inclination of the earth's axis to the ecliptic; and let S be a lighted candle placed in the centre of the frame, representing the sun in the middle of the planetary system; with the ecliptic divided into signs, and the corresponding month marked against each.

If a small globe, or terella, representing the earth, be suspended by a string from E, or Libra, where the circles intersect each other, and the eye be placed a little above the light in the centre, the hemisphere of the globe will be illuminated, including both poles; the apparent situation of the sun will be in the opposite side of the ecliptic F, or in the sign Aries; and this is called the vernal equinox, which happens about the 20th of March, when the terminator passes through the poles, and makes the days and nights equal all over the earth. Whilst the terrella is moving through Libra, Scorpio and Sagittarius, the terminator, or edge of the light, keeps increasing beyond the upper or north pole, where it has attained its greatest distance; then the globe is in Capricornus G, and the sun is apparently in the opposite sign Cancer H, which happens on the 21st of June. Now all the parallels of latitude in the northern hemisphere have the greatest part of their circles illuminated, and their days are the longest; but the duration of light, or the length of day, is in proportion to the distance of the parallels from the equator, increasing from it to the north pole, and decreasing in like ratio from the equator to the south pole.

But with respect to the terrella: Whilst we trace the return of the terminator towards the north pole through Capricornus, Aquarius and Pisces, we perceive it advance towards the south, till the terrella is in the beginning of Aries, and the apparent place of the sun is in Libra; then the terminator passes through the poles, and the days and nights are again of an equal length, which happens at the autumnal equinox, about the 22d of September. Let the terrella be moved on through Aries, Taurus and Gemini, till it reach the first degree in Cancer, and the sun's apparent place will be in Capricornus; during its progress through these signs, the illuminated part, or the terminator, leaves the northern pole in darkness, and enlightens the regions about the southern. Now the greater part of the circles or parallels of latitude in the northern hemisphere are in darkness, whilst those in the southern have their greatest portion of light, which produces mid-winter to the former, and midsummer to the latter; and this takes place about the 21st of December. By moving the globe through the three remaining signs, Cancer, Leo and Virgo, the terminator again approaches towards the north, illuminates both poles, regains its first position in Libra, and completes its annual revolution.

The change of seasons is represented by a very simple apparatus, consisting of a wire orbit about four feet in diameter, and a six inch globe. The wire may be suspended vertically near a wall, with a candle in the centre. The globe must be sawed into two parts in the line of the ecliptic, and held together by an iron axis, with a screw at each end. The globe may be carried around on the wire, screwing it fast at every twelfth division; and the illuminated part will represent the season for each month.

What is usually called the summer, that is, from the vernal till the autumnal equinox, is nearly eight days longer than from the autumnal till the vernal; for the sun, in passing through the six northern signs, Aries, Taurus, Gemini, Cancer, Leo and Virgo, performs its apparent motion in 186d. 11h. 51m.; but in passing through the winter signs of Libra, Scorpio, Sagittarius, Capricornus, Aquarius and Pisces, it only takes up 178d. 17h. 58m. which makes a difference of 7d. 17h. 53m.

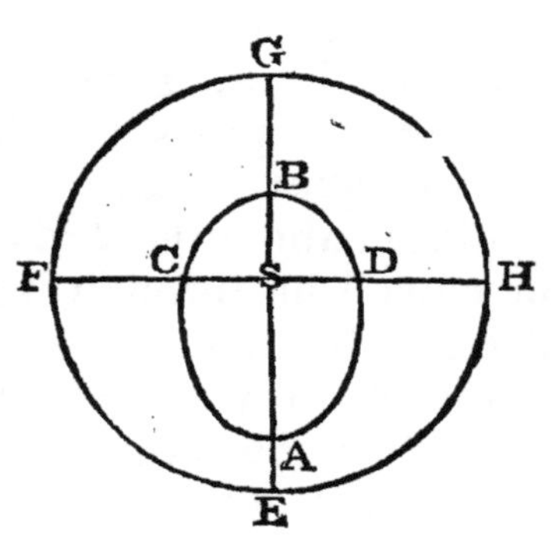

Let A B C D represent the earth's orbit, and S the sun in one of its foci; when the earth is at C, the sun appears at H the first sign Aries, and as the earth moves through C B to D, the six southern signs, the sun appears to move through the six northern H E F. In like manner, whilst the earth passes through the northern signs D A C, the sun passes through the southern F G H, the corresponding circle in the heavens to half of the earth's orbit C B D. Thus, the line F H divides the ecliptic into two equal parts, and the elliptical orbit of the earth into two unequal parts; the greater part C A D is that which the earth describes in summer, and the less is its winter course. Beside these unequal divisions of the earth's orbit, it apparently moves slower in its aphelion, or the distant part of its course, than in its perihelion, or when it is nearest the sun.

The apparent diameter of the sun is greater in

winter than in summer; for the sun is considerably nearer the earth whilst the earth passes through the winter signs C B D, than whilst it passes through the summer signs D A C.

It may naturally be asked, why the winter is colder than the summer, as the sun is nearer the earth? In summer, as before observed, the sun rises much higher above the horizon; therefore, its rays fall in a greater quantity, and more directly upon the earth than in winter: likewise the length of day, or the time that the sun is above the horizon in summer, being much longer than in winter, the earth and atmosphere receive more heat in the day than they lose in the night, so that we have a gradual accumulation of heat during the summer months, which makes it generally hotter after the sun has passed the summer solstice, or tropic of Cancer, than in any of the preceding signs.

The *Moon* is one of those heavenly bodies which we call satellites: it is secondary to the earth, and revolves round it, whilst the earth performs its annual course.

The moon's apparent place, viewed by a spectator on the earth, is extended to a great circle of the heavens, and seems to move through the twelve signs of the zodiac in a month, or lunar day.

The plane of the moon's orbit, if it were extended, would intersect the ecliptic in two points, making an angle with it of about five degrees; but this inclination varies, being greatest when the moon is in its quadrature; and least when it is in conjunction or opposition with the sun.

The two points where the moon's orbit cuts the ecliptic are called the nodes: when the moon ascends from the south to the north side of the ecliptic, it is called the ascending node, and from the north to the south, the descending node. The line of the nodes is not always directed to the same point, but has a motion contrary to the order of the signs, and by this retrograde course, it completes its circuit in 18y. 225d. at which time the line of the nodes returns to the same point in the ecliptic.

When the moon crosses the ecliptic, it is in its nodes; but in all other parts of its orbit it is in north or south latitude, according as it is above or below the ecliptic.

The mean time of a revolution of the moon about the earth, that is, from one new moon to another, is called a synodical month, or lunation, and consists of 29d. 12h. 44m. The line of its revolution round the earth, from any point in the zodiac to the same point again, is called a periodical month, and contains 27d. 7h. 43m. The moon moves in its orbit about 2290 miles in an hour, and only turns once round its own axis whilst it makes a revolution round the earth, which causes it always to present the same side towards the earth, and makes its day and night of the same length as our lunar month.

If the earth were stationary, the periodical and synodical months would be the same; but as the earth keeps moving forwards in its orbit, whilst the moon is performing its revolution, it has not only to pass through its own orbit, but has likewise to overtake the earth again in its passage through the ecliptic.

For if S be the sun in the centre of the system, E O part of the earth's orbit, and M A the orbit of the moon; when the moon is in conjunction at A, if the earth remained at B, whilst it made its revolution A, *a*, M, *b*, A, the periodical and synodical months would be the same: but, during this revolution, the earth has passed on in its orbit to C; therefore the moon must advance to C before the earth and moon can come into conjunction again; but it is obvious by inspection, that when the moon has arrived at F, it will have completed its revolution round its orbit, and the time of performing the remaining arc F *c*, will be the difference of time between the periodical and synodical month, which is about two days and five hours.

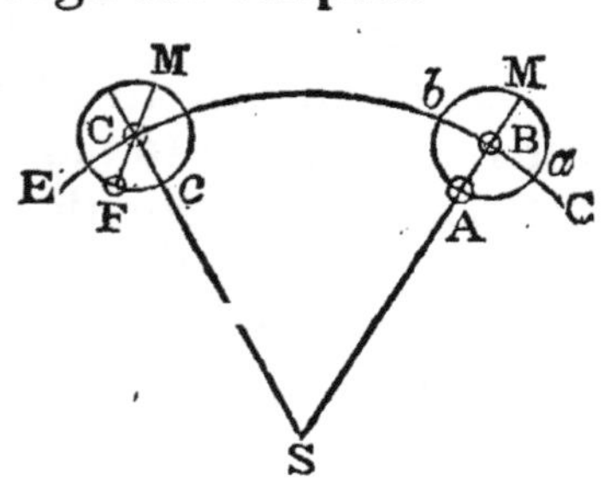

Phases of the Moon.

The moon is a dark opake body, moving round the earth in a small orbit, and shines by a borrowed light from the sun, which illuminates one half of its body, and leaves the other in darkness. We perceive different degrees of this illumination, according to the various positions of the moon with respect to the sun and the earth: hence we see one half of its body enlightened, or a full face, when it is placed in opposition, or in that part of its orbit which is the most remote from the sun. When the moon is in conjunction, or in that part of its orbit which is between the earth and the sun, its enlightened face is turned from us, which renders it invisible; this is the time of new moon. When the moon appears in the intermediate part of its orbit, between the conjunction and opposition, it is in its quadratures, and about half of its illuminated surface is turned towards us. Its phases and appearances are particularly explained by the figure.

Let S represent the sun, K the earth, A B C D, &c.

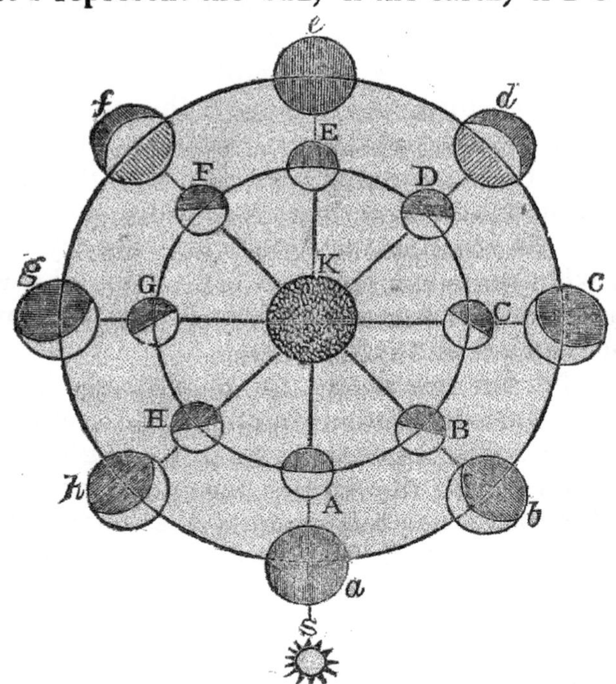

the moon in its orbit, with the sun's rays falling on that half of its surface which is opposite to the sun; and the outer circle *a, b, c, d,* &c. the various phases of the moon, as they appear to a spectator on the earth, during the whole of a lunar month.

When the moon is in conjunction with the sun at A, the darkened side is presented towards the earth; therefore, being an opake body, it becomes invisible, and it is then called the time of new moon. When it has passed an eighth of its orbit, and arrived at B, a quarter of its enlightened surface will be turned towards the earth, and it will appear horned, as at *b*. In passing another eighth of its orbit, it arrives at C, and it is then in its quadrature; one half of its surface appeared illuminated at *c*. After having passed on to D, it appears gibbous, and more than one half of its face is illuminated, as appears at *d*. At E, the whole face is seen bright, which is called opposition, or the full moon. Thus, after having attained its fullest appearance, it again begins to decline from E to F, and appears gibbous at *f*: thence it passes to G its quadrature, and is seen at *g*, half illuminated; then, after being horned at *h*, it completes its revolution, and falls into conjunction at A with the sun.

The earth serves to enlighten the moon, in the same manner as the moon enlightens us; but its appearance must be much larger than that of the moon to the earth, and the changes take place in contrary order; that is, when the moon appears full to us, the earth must be in conjunction with the sun, which turns the darkened surface to the lunarians.

Soon after the new moon, the whole body is dimly seen, independently of the illuminated crescent on its outer surface, which proceeds from the light that is reflected on it from the earth; for at our new moon, the earth appears as a full moon to the lunarians, and part of the light which they receive from us is again reflected back to the earth.

An eclipse of the moon is a privation of light, caused by the interposition of the earth directly be-

tween the sun and the moon. Or it may be considered as proceeding from the conical shadow of the earth, when the moon enters between the base and the vertex. As the earth's orbit is in the plane of the ecliptic when it is viewed from the sun, it is evident that the earth's shadow must tend directly to that part of the heavens; and as the moon's orbit has an inclination of about five degrees with the ecliptic, and only crosses it in two points called its nodes, the shadow of the earth cannot fall upon the moon except it is in or near one of its nodes.

Let the line A D represent a part of the ecliptic, the plane of which coincides with that of the earth's orbit, and C B part of the orbit of the moon, crossing the

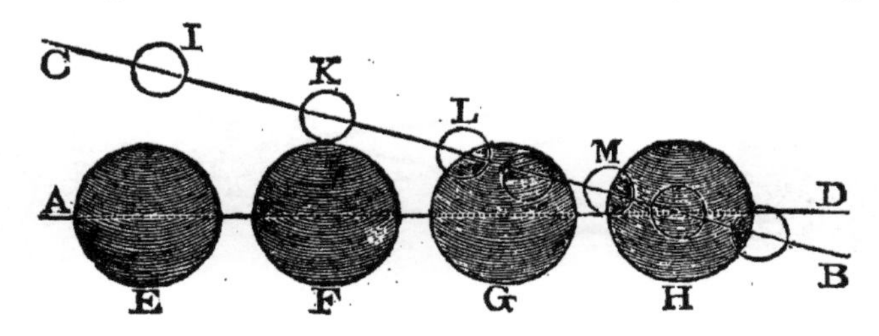

ecliptic at H, which is called its node. Then, if E F G H represent the earth or its shadow in four different positions in its orbit, when the moon I approaches its node H, and the shadow of the earth is at E, it has no part of the sun's rays intercepted; but if the earth be at F and the moon at K, a small obscuration takes place, and the moon is partially eclipsed. When the moon is at L and the shadow at G, it enters wholly into it, which is called a total eclipse; but when the moon's centre passes through the centre of the earth's shadow, as at H, which can only happen when the moon is directly in one of its nodes, it is called a total and central eclipse.

The duration of a central eclipse, or the time that the moon takes from entering the shadow to quitting it, is about four hours; during two hours of this time, the moon passes through three times the length of its diameter totally eclipsed.

The moon's diameter is supposed to be divided into twelve equal parts called digits, and the magnitude of a partial eclipse is denominated by the number of parts that are obscured; thus, if the shadow pass through a quarter of the moon's diameter, it has three digits eclipsed.

The earth, like all other opake globular bodies which receive the sun's rays, not only throws a dark converging or conical shadow behind it, but has likewise a thin diverging shadow on each of its sides, called the penumbra, which is occasioned by a partial obscuration of light from the sun.

For if S be the sun, and E the earth receiving its rays on its surface, there will be a dark shadow or

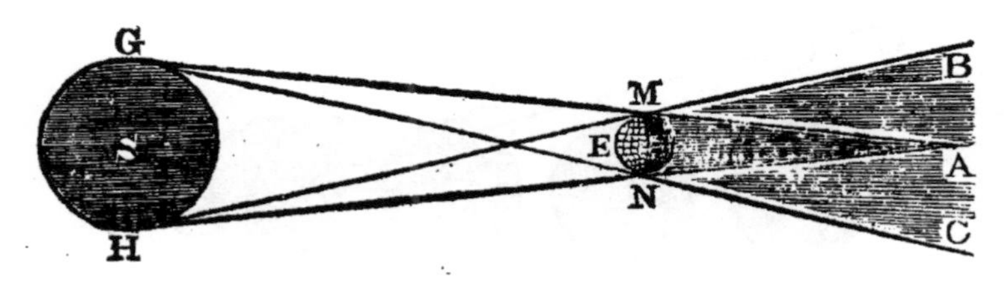

total obscuration in the cone M, A, N, which cannot receive a ray of light from any part of the sun; and a penumbra or thinner shade will fall on each side, in the angular parts B M A, C N A, increasing in darkness towards the sides M A, N A.

For the penumbra B M A can only receive a partial light from the upper part of the sun towards G, which keeps decreasing till it terminates at the side of the cone in the line G M A. In like manner the penumbra A N C is deprived of any rays from the upper part of the sun, and is only partially illuminated by the lower towards H, till the rays terminate in the line H N A.

The moon passes through the penumbra before it enters the dark shadow, and afterwards traverses the opposite shade before it resumes its ordinary brightness. It may be distantly perceived when it is in the outer side of the penumbra; but when it approaches near to the dark cone, its surface is much more obscured.

Lunar eclipses are visible over every part of the earth, that has the moon at that time above the horizon; and the eclipse appears of the same magnitude to all from the beginning to the end. On the northern side of the equator, the eastern side of the moon enters the western side of the shadow, and passes out by the eastern. Total central eclipses are of the longest duration; that is, when the diameter of the earth's shadow passes through the centre of the moon in its nodes; as the moon quits its nodes, either into north or south latitude, the eclipses become more partial and of less duration. The length of an eclipse, even in the nodes, is not always the same; for if it happen that the moon is in apogee, and the earth in aphelion, their greatest distances from each other, the length of the eclipse will be about 3h. 57m.; but if it take place when the moon is in perigee, and the earth in perihelion, their nearest distance, then the duration will be 3h. 37m. only.

The moon, in the midst of an eclipse, has usually a faint copperish appearance: this is supposed to proceed from the rays of light which are refracted by the earth's atmosphere, and fall upon the surface of the moon.

The moon's nodes have a motion from the consequent to the antecedent signs, which move about 19$\frac{1}{3}$ degrees in a year; so that in 18y. 225d. it passes through all the signs in the ecliptic, and returns again to the same point. If the moon's nodes were fixed in the same part of the ecliptic, there would be just half a year between the times of the sun's conjunction with the nodes; therefore, in whatever sign or month of the year an eclipse should take place, it would always happen at the same time in every succeeding year; but as the moon changes the situation of its nodes in the ecliptic for 18y. 225d. this will be the period of succession before the same eclipses fall in the same part of the ecliptic.

An eclipse of the sun is caused by the interposition of the moon between the sun and the earth. This seems more properly called an eclipse of the earth, as the sun loses no part of its brightness; but the intervention of the moon between the sun's face and the earth, causes a partial darkness upon a small part of the earth's surface, accompanied by a penumbra, or thinner shade, like that which was explained in the preceding subject.

In a solar eclipse, let S represent the sun, L the earth, and M the moon in that part of its orbit which

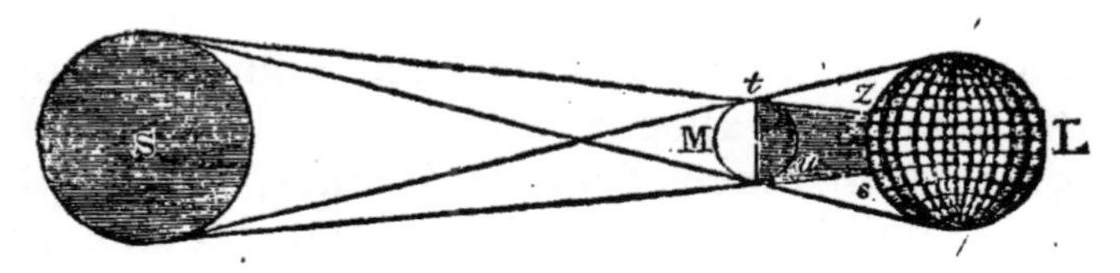

is called in its conjunction, or between the earth and the sun; then *z, s, t, u,* is the moon's conical shadow, which passes over *z, s,* a small part of the earth, and produces an eclipse, or withholds the sun's rays from that part of its surface; and on each side of the conical shadow is the penumbra, or shade, which is caused by the partial deprivation of the rays of the sun.

Now as the moon is so much less than the earth, it can only cover a small part by its shadow; therefore those parts that are out of its shade can perceive no appearance of an eclipse: even a central eclipse, that is, when the moon's centre passes through the diameter of the sun, can only be visible to all those who have the moon above the horizon.

In solar eclipses, the moon's shadow upon the surface of the earth does not, in general, exceed 180 miles in diameter; though the penumbra extends several hundred miles round. If the eclipse happen when the moon is exactly in its nodes, it will cast a circular shadow on the earth; but when the moon has northing or southing, the shade is elliptical.

The course of the moon's shadow on the earth is generally from east to west, inclining towards the north, if it be in its ascending node, and towards the south in descending.

The whole time that the shadow and penumbra take to pass any given point, is called the general eclipse; the total eclipse is only whilst the darkest part passes the place.

In solar eclipses, the face of the moon appears covered with a faint light, which is attributed to the reflection of the illuminated parts of the earth. When the moon changes in its apogee, or greatest distance from the earth, its shadow is not sufficiently long to reach to its surface, and the sun appears like a luminous ring round the dark body of the moon, and forms what is called an annular eclipse.

Tides.

The moon, by the attraction of gravitation, tends to lengthen the earth and its atmosphere in a line perpendicular to its surface, and thereby causes tides in the ocean.

Suppose the earth were an entire globe of water, the moon would attract all of the mass towards itself; but the diameter being about 8000 miles, the side nearest the moon would be most attracted; consequently, the part next to the moon would be drawn a certain distance, the middle less, and the opposite part still less. This would raise the water nearest to the moon by excessive attraction, and leave the opposite side raised by comparatively diminished attraction, while the sides would be compressed by the receding of the water towards the two raised sides. On this principle tides are produced, though the interposition of the continents prevents that regularity which would take place without them.

When the sun and moon are on the same side, or on opposite sides of the earth, they jointly tend to lengthen it in the line of their arrangement with the earth, and thereby cause spring tides.

The same principle which governs in the preceding case, applies here; but when these two powers act in concert, in and near the full and change of the moon, the tide is highest, called spring tide. In the moon's quarters, they act in opposition on the same principle, causing the lowest tides, called neap tides; for then the tides depend on the excess of the sun's attraction

Tides in the atmosphere are governed by the same laws; but the air being more easily set in motion, it obeys the sun and moon sooner: of course its tides are a little forward of those of the ocean.

EXPLANATION OF TERMS

IN

THE PRECEDING WORK.

Aberration, in optics, the deviation or dispersion of the rays of light when reflected by a speculum or refracted by a lens, by which they are prevented from meeting or uniting in the same point, and then produce a confusion of images.

Acceleration, the increasing velocity of heavy bodies as they fall towards the centre of the earth, by the force of gravity.

Achromatic telescope, a species of refracting telescope which produces the images of objects bright, distinct, and uninfected with colours about the edges, through the whole extent of a very large field or compass of view.

Analogy, the comparison of several ratios of quantities or numbers one to another.

Aphelion, that point in the orbit of a planet in which it is at its greatest distance from the sun.

Apogee, that point of the orbit of a planet which is farthest from the earth.

Apses, in astronomy, are the two points in the orbits of the planets, where they are at their greatest and least distances from the sun or the earth.

Atmosphere, a term used to signify the whole of the fluid mass, consisting of air, aqueous and other vapours, electric fluids, &c. which surrounds the earth to a considerable height.

Attenuate, to weaken or rarefy.

Attrition, the action or rubbing of one body upon another.

Caloric, supposed to be that elastic fluid which produces heat.

Capillary tubes, extremely fine tubes, like hairs.

Catoptrics, the science of reflex vision, or that part of optics which explains the laws and properties of light reflected from mirrors or specula.

Centrifugal force, is that by which a body revolving about a centre, or about another body, endeavours to recede from it.

Centripetal force, is that by which a moving body is perpetually urged towards a centre, and made to revolve in a curve instead of a right line.

Cohesion, that principle by which the particles of matter in all bodies combine and stick together.

Collapsing, falling together.

Collision, the dashing or striking together of two bodies.

Concave, an appellation used in speaking of the inner surface of hollow bodies, more especially of spherical or circular ones.

Concentric, having the same centre.

Condensation, the art of compressing or reducing a body into a less bulk or space, by which means it is rendered more dense and compact.

Congelation or *freezing*, the act of fixing the fluidity of any liquid by cold, or the application of cold bodies.

Conical, of the form of a cone or sugar loaf.

Contact, the relative state of two things that touch each other, but without cutting or entering, or where surfaces join each other, without any interstice.

Converging, tending to one point.

Convex, round or curved, and protuberant outwards, as the outside of a globular body.

Corpuscles, the minute parts or particles that constitute natural bodies.

Curvilinear, bounded by curved lines, as the circumference of a circle, ellipsis or oval, &c.

Density, that property of bodies by which they contain a certain quantity of matter under a certain bulk or magnitude.

Diagram, a scheme for the explanation or demonstration of any figure, or its properties.

Dioptrics, the doctrine of refracted vision, or that part of optics which explains the effects of light, as refracted by passing through different mediums, as air, water, glass, &c. and especially lenses.

Disk, the body or face of the sun or moon, which appears to us in a circular plane, although it is a spherical body.

Diverging, in optics, is particularly applied to rays, which, issuing from a radiant point, or having in their passage undergone a refraction or reflexion, do continually recede farther from each other.

Eccentricity, is the distances between the centres of two circles or spheres which have not the same centre, or the distance from the centre of an ellipse to one of its foci.

Effluvium, a flux or exhalation of minute particles from any body, or an emanation of subtile corpuscles from a mixed sensible body, by a kind of motion or transpiration.

Elongation, the removal of a planet to the farthest distance it can be from the sun, as it appears to an observer on the earth.

Equilibrium, equality of weight, equal balance between two forces acting in opposite directions.

Evaporation, the act of dissipating the humidity of a body in fumes or vapour.

Expansion, the swelling or increase of the bulk of a body when acted upon by a superior degree of heat, or the effect produced by rarefaction.

Ferruginous, partaking of the nature and quality of iron.

Fixity, a property which enables a body to endure fire and other violent agents.

Focus, in optics, is the point of convergency, or that where the rays meet after refraction or reflection

Friction, the rubbing together of two bodies.

Fulcrum, a fixed point about which a lever, &c. turns and moves.

Gibbous, a term applied to the moon when she appears more than half full, or enlightened, to distinguish her from the state when she is less than half full, or a crescent.

Globules, very small spherical bodies.

Gravity, weight, or that quality by which all heavy bodies tend towards the centre of the earth.

Hemisphere, half a globe or sphere.

Heterogeneous, composed of different kinds, natures, or qualities.

Horizontally, parallel to the horizon.

Horizon, a circle dividing the visible part of the earth and heavens from that which is invisible.

Hypothesis, in philosophy, denotes a kind of system laid down from our own imagination, by which to account for some phenomena or appearances of nature.

Ignite, to kindle or generate fire.

Impulsion, thrusting forwards, or driving on.

Inflexion, in optics, called also diffraction and deflection of the rays of light, is a property of them, by reason of which, when they come within a certain distance of any body, they will be either bent from or towards it, being a kind of imperfect reflection and refraction.

Intensity, the degree or rate of the power or energy of any quality.

Interstices, spaces between, or where parts are not in contact.

Lens, a piece of glass or other transparent substance, so formed that the rays of light in passing through it have their direction changed.

Maximum, the greatest quantity, force, &c. which can take place under certain circumstances.

Medium, denotes that space or region of fluid, &c. through which a body passes in its motion towards any point.

Meridian, in astronomy, is a great circle of the celestial sphere passing through the poles of the world, and both the zenith and nadir, crossing the equinoctial at right angles, and dividing the sphere into two equal parts or hemispheres, the one eastern and the other western.

Meridian, in geography, is a great circle passing through the poles of the earth, and any given place the meridian of which it is.

Momentum, the quantity of motion in a moving body.

Oblate, flatted at the poles.

Opake, dark, thick, not transparent.

Orbit, the course in which any planet moves.

Oscillation, or vibration, is the reciprocal ascent and descent of a pendulum.

Pendulum, any heavy body so suspended as that it may swing backwards and forwards about some fixed point by the force of gravity.

Penumbra, a faint or partial shade in an eclipse, observed between the perfect shadow and the full light.

Perihelion, that point in the orbit of a planet or comet which is nearest to the sun: in which sense it stands opposed to aphelion, which is the highest or most distant point from the sun.

Perigee, is that point in the heavens in which the sun or any planet is nearest to the earth.

Phases, the various appearances or quantities of illumination of the Moon, Venus, Mercury, and the other planets, by the sun.

Phenomenon, a singular appearance in nature.

Porosity, the quality of being porous, or full of small holes.

Quadrature, in astronomy, that aspect or position of the moon when [illegible] 90 degrees distant from the sun.

Radiant, any point from which rays proceed.

Ramous, full of branches or fibres.

Ratio, the rate or proportion which several quantities or numbers bear to each other.

Refrangible, capable of being refracted or bent.

Reflection of the rays of light, is their motion after being repelled or reflected from the surface of bodies.

Refraction of light, is an inflection or deviation of rays from their rectilinear course, on passing obliquely out of one medium into another of a different density.

Reflexible, capable of being reflected.

Rotatory, turning round.

Retardation, the act of retarding, that is, of delaying the motion or progress of a body, or of diminishing its velocity.

Retina, the expansion of the optic nerve.

Satellites, certain secondary planets moving round the other planets, as the moon moves round the earth.

Segment, is a part cut off the top of a figure by a line or plane: the segment of a circle is a figure contained between a chord and an arc of the same circle.

Semidiameter, a line drawn from the centre of a circle to any part of its circumference.

Sine, a right line drawn from one extremity of an arc, perpendicular to the radius, drawn to the other extremity of it.

Solstice, is the time when the sun is at the greatest distance from the equator. There are two solstices in each year, called the summer solstice and winter solstice.

Speculum, any polished body which reflects the rays of light.

Tangent, a right line drawn on the outside of a circle, and just touching its circumference.

Vacuum, a term used in philosophy to denote a space entirely devoid of all matter.

Vertex, the top of any line or figure: in astronomy, that point in the heavens immediately over our heads.

Vesicle, a small cuticle filled or inflated, or a very small bladder.

Visual rays, are lines of light conceived to come from an object to the eye.

Volatilize, to separate the particles of a body, or to make it evaporate.

THE END.

Lightning Source UK Ltd.
Milton Keynes UK
UKHW020749251118
332796UK00002B/92/P